W9-CDY-503

Messages from the Stars

Also by Ian Ridpath

WORLDS BEYOND

ENCYCLOPEDIA OF ASTRONOMY AND SPACE (editor)

MESSAGES FROM THE STARS

Communication and Contact with Extraterrestrial Life

Ian Ridpath

1817

HARPER & ROW, PUBLISHERS
New York, Hagerstown, San Francisco, London

 For information address Harper & Row, Publishers, Inc., 10 East 53rd Street, New York, N.Y. 10022.

FIRST U.S. EDITION

Designed by Eve Callahan

Library of Congress Cataloging in Publication Data

Ridpath, Ian.
Messages from the stars.
Includes index.
1. Life on other planets. I. Title.
QB54.R4 574.999 78-2160
ISBN 0-06-013589-1

---

78 79 80 81 82 10 9 8 7 6 5 4 3 2 1

To Carl Sagan

All the night
I heard the thin gnat voices cry
Star to faint star across the sky

Rupert Brooke
The Jolly Company

Contents

PART THREE: VISITORS FROM THE STARS?

Preface

Does life exist elsewhere in space? This is the single most important scientific question which we are currently capable of answering, and certainly the most exciting. *Messages from the Stars* is an attempt to tell the story of the search for life in space, a story which should appeal to every human being and particularly to those millions who buy science fiction either in the form of self-confessed adventure stories or in the guise of speculative fact as purveyed by writers on UFOs and ancient astronauts.

Messages from the Stars begins by looking at the scientific basis for the belief in life elsewhere in space, and at the attempts by scientists to find such life, either by sending spacecraft such as Viking in search of lowly life forms on the planets of our own solar system, or by listening with radio telescopes for signals from advanced civilizations on planets circling other stars. The book also considers how we might colonize space and travel to the stars, as well as speculating upon the consequences of such actions. It ends with a critical examination of the claims by those such as Erich von Daniken and writers on UFOs that we already have been, or are being, visited by extraterrestrials, and I attempt to place these fantastic claims in perspective against the results of the scientific search for extraterrestrial life.

When I put the finishing touches to my first book on this subject, *Worlds Beyond*, at the end of 1974, I did not suspect that two years later I would have embarked on a sequel. But such has been the pace of advance that this sequel has become necessary. The scientific search for extraterrestrial life has become a boom subject. It has grown from the part-time concern of a few enthusiasts to the stage at which the National Aeronautics and Space Administration has put its authority and facilities behind plans for two major new searches. Within the next five to ten years we should either have detected messages from the stars or, if not, we will at least have placed severe constraints upon the likelihood of their existence. Whether we find messages or not, I strongly suggest in Chapter 8 of this book that we should also consider whether to begin transmitting deliberate signals of our own.

If a sense of puzzlement, even of paradox, emerges partway through the book, it is an accurate reflection of the reaction of many people, myself included, to the fact that, while life *ought* to be abundant throughout space, we have as yet found no clear sign of it. As of now, it must be admitted that we do not know that there is life out there to be detected. But I for one would like to know whether or not there is life elsewhere in space, and it will certainly be exciting to try to find out. The discovery that we are not alone would rival major landmarks in our civilization's development, such as the release of atomic power and the first landing on the Moon, and I believe it would eventually come to dominate them in importance. For both author and reader, the search for extraterrestrial life is the biggest science story of our day.

Since the subject is so big, it is not surprising that it takes more than one average-sized book to do it justice. Although *Messages from the Stars* is designed to be read and understood on its own, readers interested in more background information on some of the topics covered will find this in its predecessor, *Worlds Beyond* (New York: Harper & Row, 1975). The two books are intended to be complementary.

As before, I must thank the scientists and other thinkers who have made this subject so rich in activity and insight; without them there would be no book to write. They are credited individually throughout the book wherever their results or ideas appear. I must also thank Graham Massey and Peter Spry-Leverton of BBC-TV, producer and researcher respectively of the Horizon special on ancient astronauts (televised in the United States as part of the PBS *Nova* series) for help with the last section of the book, particularly chapters 11 and 13. And to Buz Wyeth and Steve Roos, who believe in giving readers facts, not fantasy.

Ian Ridpath

PART ONE

LIFE AMONG THE STARS

1

Are We Alone?

In 1976, the United States landed two automatic biology laboratories on the surface of the planet Mars—a billion-dollar commitment to the possibility that life has arisen on another planet in space.

In 1974, a radio message announcing the existence of human life on Earth was beamed from the giant Arecibo radio telescope in Puerto Rico toward a cluster of 300,000 stars in the constellation Hercules. At present, the Pioneer 10 and 11 spacecraft are heading through the solar system before leaving the Sun's environment entirely and drifting out toward the stars, carrying plaques engraved with a message to other civilizations that might intercept them deep in the Galaxy.

By 1980, the National Aeronautics and Space Administration hopes to begin a large-scale search for possible interstellar radio messages, building on the experience of several pilot projects by radio astronomers in recent years. The Soviet Union in 1974 announced two five-year plans of its own to detect messages from the stars.

Astronomers with large telescopes are currently scanning nearby stars for the existence of planets, potential habitats of life elsewhere in the Galaxy. A group of British engineers and physicists is outlining the design for a nuclear-powered starship that could reach the nearest of these other planetary systems.

Against the backdrop of these scientific investigations, many citizens of Earth are prepared to believe that other civilizations in space are undertaking similar projects—with us as a target. Optimists have placed bets with Ladbroke's, the London bookmakers, that a UFO will land or crash on Earth with alien life on board. A woman named Ruth Norman from El Cajon, California, who claims to have been in touch with extraterrestrials, regularly renews her bet on this proposition; she has so far staked a total of several thousand pounds.*

The scientific investment in the search for extraterrestrial life is also, in its own way, a gamble, since no one has any real proof that such life will be found. But the scientific gamble is sure of at least some return, for the investigation will clarify understanding of the formation of stars, planets—and of life itself.

The popular belief in life beyond Earth frequently manifests itself as a kind of space-age religion, such as in the form of flying-saucer cults. In contrast, the scientists' belief is based on a more rational assessment of the Universe around us.

As long ago as 1961, a group of highly qualified American scientists held a symposium at the National Radio Astronomy Observatory at Green Bank, West Virginia, to estimate the possibility of extraterrestrial life. In the Soviet Union, similar informal discussions were also under way. The two sides met for an international scientific summit in 1971 at the Byurakan Astrophysical Observatory in Soviet Armenia, where they concluded that as many as one million civilizations capable of interstellar communication may exist in our Galaxy.

To arrive at their conclusion, the scientists discussed what is known as the Drake equation, named after the American radio astronomer Frank Drake, who first put it forward at the Green Bank conference in 1961. Drake is also famous as the progenitor of Project Ozma, the first attempt to listen for messages from the stars. The Drake equation allows scientists to calculate the number of civilizations in space from an assessment of the various unknowns involved in the origin and evolution of extraterrestrial life. Unfortunately, the estimates for most of the unknowns are little more than educated guesses, and the final conclusion is correspondingly uncertain. Improving our understanding of the various factors in the Drake equation is one of the main tasks in the search for life beyond Earth.

* But UFO skeptic Philip Klass of Washington, D.C., has risked $10,000 that such an event will not happen. See Chapter 16.

The Drake equation multiplies together the following factors: the rate of star formation at the time the Sun was born; the fraction of those stars with planets; the number of planets in each system that are suitable for life; the fraction on which life actually does arise; the likelihood that such life will develop to the stage of intelligence; the desire among that life form to communicate; and the longevity of the civilization. This simple formulation has been the jumping-off point for much speculation about life in space.

Each of the factors seems innocent enough in itself, but each conceals a morass of assumption and speculation. One factor that we can assess with reasonable accuracy is the first: the rate at which stars were formed when the Sun itself came into being. Stars are glowing balls of gas, and the Sun is an average example. If we could get as close to the other stars in the sky as we are to the Sun, we would see that all stars are incandescent spheres, some larger and hotter than the Sun, others smaller and cooler, but all emitting their own light and heat.

Because of their extreme distance we are unable to examine stars in the same detail as we study the Sun, but astronomers can sort stars into broad classes of size and temperature by analyzing their light in spectroscopes attached to telescopes. In this way the full range of star types from blazing supergiants to feeble dwarfs has been listed, and with the help of theorists a consistent picture of their birth, evolution, and death has been pieced together. Fortunately for us, it turns out that the Sun is a relatively long-lived star, in stable middle age, and certain to go on shining steadily for thousands of millions of years.

All the stars we see in the sky are part of a Catherine-wheel-shaped aggregation known as the Galaxy. The Sun lies about two-thirds of the way to the edge of this stellar congregation, in the galactic suburbs. The nearest stars to us appear scattered over the night sky, forming the familiar shapes of the constellations. The more distant stars of the Galaxy cannot be seen individually but form the faint, hazy band around the sky known as the Milky Way. The name Milky Way is often used as a synonym for our Galaxy.

Astronomers estimate that there are 100 billion stars or more in our Galaxy. The whole Galaxy is so large that it would take a beam of light 100,000 years to cross from one side to the other. In the jargon of astronomy, the Galaxy is therefore said to be 100,000 light-years in diameter. The distances between individual stars are of the order of a few light-years. For instance, Alpha Centauri, the nearest star to the Sun, lies 4.3 light-years away.*

The Galaxy in its early days, 10 billion or so years ago, consisted mostly of hydrogen gas from which stars were forming. As the amount of gas declined, so did the rate of star formation. The Sun is believed to have formed about 4.6 billion years ago, when the Galaxy was approximately half its present age, so the rate of star formation at that time was probably the average of that over the entire history of the Galaxy. Production of 100 billion stars over a span of 10 billion years means an average birth rate of 10 stars a year. We have therefore estimated the first factor in the Drake equation.

The second factor, the existence of planets, is more problematic. Planets are nonluminous bodies smaller than stars; a total of nine planets, including Earth, orbit the Sun. The Sun and planets are believed to have come into being together from the same cloud of dust and gas in the Galaxy, a process which theory suggests should happen around many other stars. Spotting planets around other stars is difficult, and not one other planetary system has been identified with complete certainty. However, astronomers are stepping up their search to clarify this vital factor (see Chapter 2). For the moment, it seems reasonable to suppose that at least one star in ten may have planets.

How many planets in each planetary system will possess the right conditions for life? At least one such planet exists in our own solar system—the Earth. Probably the number would be the same in other systems. But the very existence of a suitable planet is no guarantee that life really will arise; Mars seems to be an example of a planet with the right ingredients for life but on which life never gained a foothold. With no firm knowledge to go on, we have no clear idea of how likely or unlikely the emergence of life is.

My own suspicion is that while plant life is easily formed, the emergence of major forms of animal life is the most difficult evolutionary step of all. A typical example of extraterrestrial life could be a forest of trees or a meadow of tall grass. Plant life is probably dominant throughout space; it certainly dominated the existence of life on this planet, up until 600 million years ago when the first sea creatures crawled onto land.

* Strictly, Alpha Centauri is a group of three stars linked by gravity. The name when used without further qualification is taken to mean all three stars.

Some scientists believe the emergence of life is very likely on suitable planets, while others think it would be rare. Similar disputes occur in estimating how frequently life will become intelligent and develop the ability and desire to communicate with other beings. Perhaps life is commonplace, but only rarely does it become able and willing to communicate. Or perhaps life is infrequent, but the evolutionary advantage of intelligence is so great, and the desire to inquire is such an inescapable aspect of intelligence, that life will nearly always develop to an advanced state.

Whatever the truth, the general conclusion from these discussions among scientists is that the Drake equation can be boiled down to the statement that the number of advanced civilizations in the Galaxy at the moment is 1/10 L, in which L is the lifetime of each civilization—the one factor we have not so far taken into account. Without worrying about L for the moment, the conclusion means that one technological civilization like our own comes into being in the Galaxy every ten years or so. This figure could, however, be badly in error because of the uncertainties in estimating each of the individual factors.

The lifetime of each civilization is important because not all the civilizations formed in the past will still be with us: many, or all, may already have died out, which would leave us with no one to talk to. If the lifetime of each civilization is short, then our contemporaries in the Galaxy are likely to be so thinly spread that we will have to search a long way off in order to find them. But if lifetimes are long, then we could be lucky enough to have intelligent neighbors around some of the closest stars. Therefore, to complete our calculation of the total number of our contemporaries in the Galaxy, we must estimate the average lifetime of all advanced civilizations.

But here we are stuck, because we have absolutely no evidence to go on. Our own civilization has reached the advanced stage of being able to communicate with the stars in only the past decade or so. We have no way of knowing whether the mistakes of this faltering young civilization will wipe it out within the next few centuries, or whether we shall transcend our present difficulties of population, food, energy, pollution, and unruly behavior toward one another and emerge onto the stable plateau of maturity and genuinely cosmic wisdom.

One widely held view is that there are two types of civilization: the overwhelming majority which rise and then subside precipitously in only a few generations; and those that come through their bad patch to live for thousands of millions of years, which are the minority. (If all civilizations reached this stage, the Galaxy would be swarming with intelligent creatures, which we don't see.)

Optimists hope that we on Earth belong to the latter category rather than the former, but it's touch and go. Perhaps there is some subtle flaw in evolution which makes intelligent creatures so competitive that they eventually destroy themselves as well as their adversaries. The overkill capacity of nuclear weapons is the prime example of this, although a more futuristic idea was presented at the 1971 Byurakan conference by Soviet mathematician M. Y. Marov, who visualized a society of robots among whom a certain elite were given to dismantling and reassembling others in order to preserve their own power. There is no reason to suppose that intelligent robots or even advanced computers will not have the same shortcomings as mankind: ambition, greed, and envy.

Assuming that the lifetime of advanced civilizations, averaged over both long-lived and short-lived types, is 10 million years, then the number of civilizations like ours or more advanced in the Galaxy is 1 million. This works out to one civilization for every 100,000 stars, and indicates that our nearest neighbors with whom we may communicate are several hundred light-years away.

Another complicating aspect is the different life forms one might expect to find throughout the Galaxy. Noted astronomer Sir Fred Hoyle rekindled visions of his black cloud science-fiction concept in a recent paper with N. C. Wickramasinghe of University College, Cardiff. Hoyle and Wickramasinghe propose that the complex molecules of life can be made in the dust and gas clouds between the stars. The possibility here is that the cloud as a whole might eventually act as an advanced form of life, with implications that are beyond the realm of scientific speculation at present.

One limiting factor on the origin and development of advanced life forms may be the occurrence of the stellar cataclysms known as supernovae. As a giant star flares up in a supernova explosion, it throws off its shattered outer layers and sends beams of dangerous radiation flying through space. A supernova is expected within 50 light-years or so of the Earth every few hundred million years. Its consequences, as astrophysicist Malvin Ruderman of Columbia University has calculated, would be drastic for life on Earth.

Dr. Ruderman finds that radiation from the supernova could strip away the layer of ozone in our upper atmosphere that keeps out the Sun's ultraviolet light. This shielding layer could be destroyed for up to a century. Although radiation from the supernova itself need not reach the ground, the vast increase in ultraviolet light from the Sun reaching Earth would cause death of simple organisms, skin cancer in humans and animals, possible poisoning by overproduction of vitamin D in our bodies, and an epidemic of mutations. Such a catastrophe may well have happened several times in the history of the Earth. Perhaps we are lucky to have survived so far.

Whether there is anyone around for us to talk to or not, throughout the Galaxy there must be many extinct civilizations and planets inhabited by intelligent creatures who have not developed high technology. The few civilizations that rise to the level of interstellar spaceflight will thus become the biologists and archaeologists of the Galaxy, studying the various primitive life forms and remains of bygone ages on the surfaces of other planets around distant stars—the role envisaged in space-fiction stories such as the *Star Trek* series.

This aspect has been investigated by space scientists J. Freeman and M. Lampton of the University of California at Berkeley. "Throughout cosmic time the Galaxy may have harbored several hundred million technically advanced civilizations, and thousands of millions of intelligent species," they said, adding that these conclusions do not depend on the disputed conjectures about the average lifetimes of advanced civilizations.

From a modified form of the Drake equation, Freeman and Lampton calculated that the number of advanced (i.e., communicative) civilizations that have evolved so far in the history of the Galaxy is 250 million, and that the number of planets housing intelligent creatures without contact capacity has totaled ten times as many (assuming that only one in ten intelligent civilizations rises to the communicative level). Whereas we have no idea about the lifetime of an advanced civilization, we do have one very good example of the lifetime of an intelligent but noncommunicative family—our ancestors. Anthropologists say that our evolution from the stage of ape-men to the present day has taken at least 3 million years. Inserted into the Drake equation, this longevity implies that there are 3 million planets inhabited by creatures of at least apelike intelligence at present—one around every 300,000 stars, giving a distance of perhaps 150 light-years to the nearest.

Freeman and Lampton's treatment leads to the result that there are a total of 2.5 billion stars around which intelligent life once existed, making the nearest 18 light-years away; and that the remains of extinct advanced civilizations will be found around 250 million stars, meaning that we would have to search out to 40 light-years to find the nearest.

Would-be galactic archaeologists and biologists will have to wait a few centuries until interstellar travel is developed to take them on their field trips. But there is plenty to interest the rest of us in the meantime. Already we have the ability to exchange radio messages with any advanced beings that may exist around other stars, and the apparently optimistic results of the Drake equation have prompted the first brief searches for interstellar signals (see Chapter 7).

Some problems in this field, stemming from the various likely lifetimes of advanced civilizations, have been considered by Gerrit Verschuur of the Fiske Planetarium in Boulder, Colorado. Verschuur is a former radio astronomer who made one of the earliest searches for signals from other stars. Now, in effect, he admits he may have wasted his time.

If the average lifetime of a civilization really is only a century or two, then the Drake equation predicts that there can be no more than ten or twenty of them around in the Galaxy at present, making the nearest about 2,000 or more light-years distant. Even if we could find this nearest civilization among the tens of millions of stars within this radius, the travel time for a radio message each way would be 2,000 years (radio waves travel at the same speed as light), and the round-trip time for an exchange of greetings would be 4,000 years. But such a two-way conversation could never take place, because the lifetime of each civilization is too short—the line would go dead before the reply could be received. Therefore, says Verschuur, for a short lifetime "we are effectively alone in the Galaxy."

If it seems impossible for short-lived civilizations to contact each other, how much do things improve for a longer average lifetime? Verschuur is pessimistic. An average lifetime of 10 million years means a typical separation between communicative societies of 100 light-years or so, thus giving an approximate round-trip time of 200 years for messages. In a long-lived and stable society, two centuries would not be a prohibitively long time; such societies would be particularly suited to undertaking long-term projects of the sort which daunt us and our short-term mentalities.

Yet Verschuur sees difficulties, among them being the fact that two societies are unlikely to reach the same stage of development simultaneously. He illustrates his point by reference to apes: if, in thousands or millions of years, they developed the ability to talk as we do at present, the chances are that we would still be unable to exchange ideas or discuss concepts with them, because our own evolution would have progressed.

"It is simply unrealistic to believe that society on this planet, or any other planet, will be remotely similar for a thousand years—let alone a million or billion years," he declares. "The inevitable conclusion that evolution never stops leads to the assumption that L [the average lifetime of civilizations with whom we may communicate] is a small number."

This seemingly powerful analogy is weakened by many factors. Prominent among them is the fact that we already can communicate in simple form with chimpanzees: the famous chimp Washoe was taught sign language used by the deaf and dumb, and another chimp, Sarah, was taught a language using colored plastic shapes for words. The language abilities of dolphins may be even greater, but communication here is hampered by the creatures' lack of manipulative ability—perhaps the main reason why we, not they, developed technology.

Any highly advanced civilization will realize the problems of interstellar discourse and so, if it wishes to talk with as many people as possible, it will use the simplest possible method for establishing that all-important first contact. The main interest of an old and advanced civilization may be in the developing creatures of the Galaxy, as the anthropologists of the Western world are increasingly fascinated by the Stone Age societies that remain on Earth.

An altogether more optimistic outlook was presented in 1975 by Bernard Oliver of the Hewlett-Packard electronics company of Palo Alto, California. Oliver's treatment of the Drake equation and the factors affecting interstellar communication is probably the most thorough yet formulated, and is worth relating in detail.

Oliver draws attention to a factor previously overlooked: that communication might begin between civilizations that were coincidentally close together, and then spread outward to more isolated communities. He notes that the older stars near the Galaxy's center are two or three times closer together than the younger stars in the Sun's vicinity, so that interstellar communication is most likely to have begun long ago in these dense central regions.

Stars are moving through space at different speeds as they orbit the Galaxy. Therefore a star's closest neighbors are continually changing, so that stars undergo a general mixing which can only increase the number of encounters between civilizations. Additionally, Oliver points to some theoretical predictions that planets could also exist in systems where two stars orbit each other; this factor is not normally taken into account in the Drake equation, which concentrates on single stars only. Stable orbits for planets are theoretically possible around twin stars at two and a half times or more the stars' separation; in other words, the planet would orbit around the outside of both stars. In such a situation the planet would probably be too far away from the stars to be warm enough for life. But things improve if the two members of a double-star system are sufficiently wide apart for planets to exist around each individual star. Stable orbits are possible around the more massive component at up to one-half the stars' separation, while around the smaller component planets could orbit safely at up to one-quarter the stars' separation. Examples of both types of orbit are provided in the solar system: if Jupiter is regarded as a dark companion of the Sun, then the outer planets beyond Jupiter can be said to orbit both the Sun and Jupiter, while the inner planets and the satellites of Jupiter follow stable orbits around each individual component.

Observations by astronomers tend to confirm these theoretical speculations about the possible existence of planets in double-star systems (see Chapter 2). Since double stars far outnumber single stars, even though only a fraction of them will be suitable for life, Oliver estimates that this modification of the Drake equation leads to a tripling of the number of possible civilizations in the Galaxy.

Occasionally life would arise on planets orbiting both components of a double-star system. Since both members of a double-star system are of similar age, this tends to synchronize the emergence of life in their vicinity. If the average lifetime of civilizations is 10 million years, Oliver suggests that something like 100,000 such civilizations will have arisen within the same double-star system, and 1 million within a distance of 10 light-years of each other. "Interstellar communication will have been established independently countless times," he asserts.

Over the millennia, such contacts will have profoundly modified the science, philosophy, and destinies of the races in contact. Additionally, his figures suggest that there have been approximately 40 million cases where a space voyage by an advanced culture around one component of a double-star system has found either simple life or the remnants of a long-dead advanced civilization on the planets of the other component star.

Even if the lifetimes of civilizations are far shorter than the 10 million years assumed here, and there is little direct effort at sending messages to other stars, one civilization could still detect another from the radio noise accidentally leaking into space from its domestic transmissions. "For a cost less than that of space programs we have already undertaken, we could construct a receiving system that would detect the presence of signals like our own UHF television at a 50-light-year range," Oliver said. He was here referring to the Project Cyclops plan for a giant radio receiving array, described in Chapter 8.

We have been radiating UHF TV signals for fifteen years, and are likely to continue doing so for at least another thirty-five years. Even if a civilization radiates such leakage signals for only fifty years, Oliver estimates that up to 500 civilizations could have come within a 50-light-year eavesdropping range of their neighbors over the Galaxy's history.

Once such evidence of another civilization came to light, Oliver envisages, the eavesdropping society would be encouraged to search deeper into space for further signs of life and also to radiate a beacon that would attract others. With a beacon operating in the Galaxy, the possibility of interstellar contact brightens enormously. Explains Oliver: "The same system that could detect leakage radiation at 50 light-years could detect a 250-megawatt beacon at 500 light-years and thus reach one thousand times as many stars.

"The number of close civilizations, although uncertain, seems large enough to support the hypothesis that an intercommunicating galactic community of advanced cultures already exists."

Oliver's work, like that of many before him, emphasizes above all that the only way to find out for sure whether or not there are messages from the stars is to listen for them. This point was first made in 1959 by physicists Philip Morrison and Giuseppe Cocconi in a paper in which they openly suggested a search by radio astronomers for interstellar messages, and thus brought the subject out into the light; the point remains as strong today as ever.

The Drake equation cannot give us final answers, but it can act as a guide. Its most valuable contribution is that it gives us a rational basis for the hope that we are not alone in space. And, of course, any other intelligent civilization can make a similar calculation that will reveal to them both the prospects and the problems of interstellar communication. If they are by nature pessimists, they may abandon the project immediately, but one hopes they will be optimistic enough to make allowances for the problems and proceed with the business of getting in touch with whoever may be around.

There may be humane reasons why advanced beings should wish to attract the attention of other civilizations in space. If it is true that the majority of cultures live for only a short time, there may come a take-off point at which a civilization lives just long enough to exchange messages with beings elsewhere. This exchange could provide the spur that a civilization needs to enter the long-lasting phase of its evolution. Our work today on interstellar communication could be our life insurance for the future.

Ultimately, the value of our present discussions about interstellar communication may be philosophical, rather than scientific. In debating the possible lifetimes of civilizations, we become more aware of our own mortality. We should therefore strive to ensure that we are the one civilization in a hundred—or the first one ever—that survives for millions of years.

And if we are the first, we inherit a responsibility that goes far beyond the bounds of tiny Earth. For if no one before has reached the plateau of cosmic maturity, our radio call signs in the future may be the lifeline that pulls other developing civilizations in space out of the despair of their own isolation.

REFERENCES

Freeman, J., and M. Lampton, *Icarus*, Vol. 25 (1975), p. 368.
Hoyle, F., and N. C. Wickramasinghe, *Nature*, Vol. 266 (1977), p. 241.
Oliver, B., *Icarus*, Vol. 25 (1975), p. 360.
Ruderman, M. A., *Science*, Vol. 184 (1974), p. 1079.
Verschuur, G., *Astronomy*, December 1975, p. 47.

2

Stars and Planets

Every star in the sky is a sun—a potential center for life in space. Stars are born from giant clouds of gas and dust, like the famous nebula in the constellation Orion. Gravity compresses parts of the clouds into tiny spinning gas blobs called protostars, at whose centers nuclear reactions begin to flare, making the object a true star. These nuclear reactions in the star's interior produce heat and light, so that the star shines. Starlight is intimately linked with life in space.

The heat and light from the Sun is the basic source of energy that sustains life on Earth: plants use sunlight for photosynthesis, the process by which they break down carbon dioxide from the atmosphere, taking in carbon for their bodies and releasing oxygen for us to breathe. Plant life, such as plankton in the sea or grass on land, provides the first link in the food chain that leads eventually to man. Whereas an advanced technological society could provide its own energy sources and inhabit the most inhospitable depths of space, the inert chemicals from which life formed on Earth 3.5 billion or so years ago had no such ability. They had to rely on the energy of sunlight to assemble them into the first living things, which themselves continued to rely on solar energy for photosynthesis. It is difficult indeed to imagine how life could emerge and develop without the natural energy source of a nearby star.

The Sun is an ideal star to nourish life, for it is not so hot as to produce searing amounts of dangerous short-wavelength radiation, nor is it so faint that it has little warming effect on the surrounding planets. Its surface temperature is 5,800°K, which places it just below midway in the range from the coolest known stars to the hottest. In color, the Sun appears yellowish, whereas the hottest stars are blue and the coolest are red. The Sun is a stable-burning star, not given to unpredictable changes in energy output, and it is also fairly long-lived: its total predicted lifetime is about 10 billion years, of which it has completed approximately half.

A star's lifetime depends on its mass: the heaviest stars burn the brightest, using up their nuclear fuel in the shortest time. For instance, if the Sun had had only 25 percent more mass, it would already have burned out. A star such as the white-colored Sirius, with a surface temperature of about 10,000°K, is more than twice as massive as the Sun and thus is expected to live for perhaps one-tenth as long. This is not long enough for life to develop on any planets the star may have, if the pace of evolution on our own Earth is typical. By contrast, a red-dwarf star such as Barnard's star, with a surface temperature only about 2,800°K and a mass perhaps one-tenth that of the Sun, could live for 100 billion years or more, at least ten times as long as the Sun.

Therefore the hottest stars, with their intense radiation and short lifetimes, seem the least likely centers of extraterrestrial life, whereas the dimmest stars, despite their glowworm output, have the advantage of longevity to compensate. We live around a star which seems to represent a good compromise between the conflicting requirements of luminosity and longevity. Over 5 percent of the stars in the Galaxy have roughly the same mass and temperature as the Sun; the most abundant stars of all are the red dwarfs, which are ten times as frequent as Sun-type stars.

But the existence of a suitable star is not by itself sufficient for the origin of life. Astronomers have found that the gas clouds around forming stars are rich in the chemicals essential to forming life—simple substances such as carbon, nitrogen, and water. What is needed are places where these chemicals could become concentrated and give life its first footing. Planets can act as rendezvous points for the chemicals of life. Planets provide an atmosphere which traps heat, supplies chemicals, and provides pressure that keeps substances such as water in liquid form. Therefore, if we want to find potential homes of life in space, we should look for planets. So how many stars have planetary systems?

The majority of stars in the sky seem to come in groups of two, three, or even more. For example, the nearest star of all, Alpha Centauri, is actually triple, and the bright star Castor in Gemini is a system of six stars, although in both cases only one star is visible to the naked eye. A major lesson from contemporary astrophysics is that stars never seem to come into being singly; the formation of stars from clouds of gas and dust in the Galaxy seems to lead naturally to the formation of multiple stars, or of a star with planets, like the Sun and its solar system of which we are a part. Therefore planets could, in theory, be as numerous in the Galaxy as are stars. But the problem is that no planets of other stars are directly visible from Earth. Since planets by their nature are cold, nonluminous objects that shine only by reflecting sunlight and give out no light of their own, they are too faint to be seen by existing methods. Even from the nearest star, our own solar system would be invisible through an ordinary telescope.

Astronomers Helmut Abt and Saul Levy of the Kitt Peak National Observatory in Arizona have attempted to estimate the number of planet-bearing stars from a study of a number of double and multiple stars near the Sun. Their work, published in 1976, is a major advance in our understanding of the formation of stars and planetary systems, and is worth reporting in detail.

Abt and Levy examined 123 bright stars near the Sun with masses ranging from the same as the Sun to 50 percent greater—in other words, stars as bright and hot as the Sun or slightly above. In addition to sixty-nine stars already known to be double or multiple, they found another twenty-five hitherto unrecognized double stars. These discoveries were made by studying the visible stars' light. Regular variations in the wavelength of light from a star reveal that it is moving around in orbit with another star too close to be seen separately; a system of this kind is known as a spectroscopic binary.

The Kitt Peak astronomers made allowances for possible companion stars that they may have missed detecting. For instance, if the two stars orbit each other in a plane perpendicular to our line of sight, so that they do not move alternately toward and away from us, there will be no wavelength change in the spectrum and their binary nature will escape detection; or the companion star may be of such low mass that it has scarcely any effect on the movement of the visible star. Other, more distant, companion stars which ought in theory to be seen separately may also have been missed because they are too faint to pick out from the mass of background stars.

Abt and Levy made the important discovery that double-star systems fall into two distinct classes: those with orbital periods of more than about 100 years, and those which orbit more quickly. The period of a binary star's orbit, like the period of a planet around the Sun, depends on the separation of the two objects, with the widest pairs having the longest periods.

In the short-period binaries one finds progressively fewer pairs as one looks for companions with progressively smaller masses than the primary; but in the long-period systems, there are progressively more pairs as one moves to companions of decreasing mass. Abt and Levy believe that this distinction arises from the different ways in which wide-separation and close binaries are formed. Probably, the close binaries were formed when a condensing protostar split into two because it was spinning too quickly to maintain stability; the tendency is for a protostar to split into two equal halves. The wide binaries formed when two stars born from independent protostars were then captured by each other's gravity. There is no preference for equal-sized partners in this group; rather, it reflects the increased abundance of smaller stars.

The orbital period of roughly 100 years which divides these two classes of binaries corresponds to a separation between the stars about equal to the distance of the planet Neptune from the Sun, which is probably the characteristic size at which a shrinking protostar begins to divide. A mass of 2.4 times that of the Sun spread over this diameter gives a density of gas similar to that assumed for the gas cloud around the Sun from which our own planetary system formed. This fits in well with theories of planetary formation, according to which the solar system formed from a small fraction of the so-called solar nebula surrounding the Sun, the unused remainder being lost into space. Abt and Levy therefore speculate that the formation of a planetary system is one result of the splitting of a protostar.

Such a view is backed up by the fact that the gas in protostellar blobs is observed to be moving at about 1 km/sec, but as the protostar gets smaller it spins faster. By the time the blob has condensed to the size of a star, its rotation will have speeded up so that it should in theory be spinning at up to 1,000 km/sec, which is too high for stability. In fact, the observed rotation speeds of stars are about 1 to 10 km/sec. So what has happened to the lost rotation? In typical binary systems, between 100 and 1,000 times as much rotational energy is contained in the stars' orbital motions as in their axial spin; similarly, most of the rotational energy in the solar system seems to be contained in the orbital motions of the planets rather than in the spin of the Sun. This suggests that to get rid of their surplus spin, all condensing protostars split to form either a double star or a star with planets.

Abt and Levy found that two-thirds of the stars they studied had short-period stellar companions. The other one-third of stars had no detectable stellar companions. If this remaining one-third of the stars sampled were also formed by shrinking and splitting gas blobs, then they too should have companions—but these companions would be too small to be visible stars.

This brings up the interesting question of what constitutes a true star. According to calculations by the University of Virginia astrophysicist Shiv Kumar, the smallest mass at which a protostar can turn on nuclear reactions to become a genuine star is 0.07 that of the Sun, or about 70 times the mass of the planet Jupiter. Then there comes a transition region between true stars and planets. A shrinking gas blob with a mass in this region will glow from the energy of its contraction before finally fading to become a black dwarf, sometimes termed a degenerate star. Even our own Jupiter is giving out a vestige of internal heat, one cause of which might be a slight shrinkage; Jupiter has been termed the star that failed.*

According to Abt and Levy's calculations, 15 percent of their sam-

* The stars of smallest mass were once believed to be the components of the binary star UV Ceti (also cataloged as L 726-8), each of which was estimated to have a mass 0.04 that of the Sun; this would mean that they were still contracting before fading out to become black dwarfs. However, more recent work (Harrington and Behall, *Astronomical Journal*, Vol. 78 [1973], p. 1096) gives a more accurate figure of 0.1 solar mass for each object, thus making them true stars. The stars now of smallest known mass are the two components of Wolf 424 and the smaller component of the double star Ross 614; in each case the estimated mass is 0.06 that of the Sun, thus placing them on the theoretical borderline between real and degenerate stars.

ple stars should be accompanied by degenerate stars and 20 percent by planets, thus underlining the theoretical speculation that no star ever forms alone. In fact, if astronomers are right in their understanding of how stars form, each shrinking blob *has* to split in two because of its high speed of rotation. Abt and Levy also find that 72 percent of the stars in their sample are wide binaries—that is, with companions that orbit in more than 100 years. Each member of a wide binary should itself have a close twin, thus producing a quadruple system.

These conclusions have deep implications for our understanding o stars and their formation. From such a limited sample, it is difficult t be certain to how many stars Abt and Levy's findings apply. But seems that they should hold for all stars of suitably long lifetimes fo life to develop—that is, of 1.5 solar masses and smaller. Therefor around all stars in this range that are not close binaries, including th individual members of wide double-star systems, we should expect find planets on which life might conceivably form.

The conclusion helps clarify a long-standing disagreement abo whether or not planets would be expected in multiple-star system Some astronomers, such as Tom Heppenheimer of the Max Plan Institute of Nuclear Physics in West Germany, have argued that presence of a degenerate star would prevent the formation of ter trial-type planets. Robert Hohlfeld and Yervant Terzian of Cor University have suggested that the difficulties in the formation planets in a double-star system should lead to a reduction by a fa of 50 in the number of civilizations predicted by the Drake equati

But theoretical work indicates that planets, if they can form, sh be able to survive around the components of a double star, as lon the components are sufficiently separated; there is an analogy ir solar system, where Jupiter is sufficiently far from the Sun to poss family of moons like a small-scale solar system. If the component are too close, any planets would be forced out of the system o one of the stars by gravitational effects; the analogy in the solar s is Mercury and Venus, neither of which can possess moons.

The stability of planetary orbits in a binary system has been ined most recently by Robert S. Harrington of the U.S. Naval vatory at Washington, D.C. He has affirmed that planets can they are sufficiently close to the individual members of a binar

they are sufficiently distant that they orbit around the outside of both stars (the analogy here is Uranus, Neptune, and Pluto, which can be said to orbit around the Sun-Jupiter double system).

Harrington mathematically tested known binaries to see if planets could exist in them. He found that planets are stable as long as the outer body (be it a star or planet) comes no closer than three to four times the separation of the two inner objects. Additionally, the region of planetary stability for most known binaries includes the area in which conditions on the planets would be suitable for life. Interestingly, Harrington found that if the Sun were replaced by a close binary of two 0.5-solar-mass stars, or if Jupiter were replaced by a 1-solar-mass star, the Earth's orbit would remain stable and temperatures on Earth would not be significantly different from what they are today.

Alpha Centauri could be an excellent example of a multiple-star system with planets. The two main stars of Alpha Centauri are similar in nature to the Sun, and orbit each other every eighty years; presumably they are the two halves of a protostar that split. But a third member of the Alpha Centauri system is a faint red dwarf named Proxima Centauri, which is marginally the closest of the three to the Sun. Proxima Centauri is so far separated from the two main members of the Alpha Centauri system that it must take about a million years to orbit them. It was presumably captured by the other two stars, and is certainly far enough from them to possess a planetary system. Unfortunately, it has not been studied for any evidence of planets.

Richard Isaacman and Carl Sagan of Cornell University have simulated on a computer the formation of planets from a nebula around a star. Their results broadly support the observations of Abt and Levy, in that they find the production of either a multiple-star system or of a planetary system around the parent star. By varying the mass of the original cloud and the conditions within it, Isaacman and Sagan created a whole range of possible outcomes, from multiple stars with planets to a single star with a large family of asteroids. In many cases recognizable planetary systems were formed, with an average of about ten planets. Isaacman and Sagan conclude that planetary systems are most likely widespread throughout the Galaxy, although in appearances they are quite diverse—many of them will not look like our solar system.

Astronomers are now finding what they believe to be examples of planetary formation in action. In 1975 University of Minnesota astronomer Edward Ney reported an hourglass-shaped blob of gas, dubbed the Egg nebula, in the constellation Cygnus. Its shape is apparently caused by a ring of dark dust around the nebula's central star, which absorbs its light to produce the dark "neck" of the hourglass, and re-radiates the star's energy at infrared wavelengths. One possibility is that the ring of obscuring dust is similar to the dense disk around the young Sun from which our own planetary system is believed to have formed 4.6 billion years ago. In 1977 astronomers at the University of Arizona, led by Rodger I. Thompson, announced the discovery of a forming star in Cygnus, designated MWC 349, surrounded by a flat disk from which planets may be forming.

Other examples of stars surrounded by obscuring rings of material include the so-called double-fan nebula in northeastern Orion; the two bright "fans" are caused by a young star illuminating a surrounding cloud of gas, with a dark lane between them which is apparently the shadow of a disk of dust around the star's equator.

Astronomer George Herbig of Lick Observatory (on Mount Hamilton near San Jose, California) has been foremost in examining young stars, known as T Tauri stars after their prototype, which still seem to be settling down after their recent birth from gas blobs. T Tauri–type stars appear to be ejecting vast amounts of matter, as is believed to have happened when the remaining nebula around the young Sun was swept away, leaving only the planets.

T Tauri stars should provide considerable information about planetary formation in action, and a group of Swedish astronomers, led by G. F. Gahm of Stockholm Observatory, believe they have found just such an example of a forming planetary system around the star known as RU Lupi, in the southern hemisphere constellation of Lupus, the Wolf. Observing from the European Southern Observatory in Chile, the Swedish astronomers found that the star's brightness rises and falls every few days, apparently due to the passage of obscuring material across the star's face. They conclude that these obscuring bodies are actually dust clouds having the mass of asteroids and moving at planetary distances—in other words, mini-planets in the process of formation. This fits very well with the assumptions of theorists that the disk of dust around the Sun first built up asteroid-sized bodies which then aggregated into planets. Remnants from this sweeping-up process are the meteorites that occasionally hit Earth.

A rather different example of possible planetary formation is provided by the star Epsilon Aurigae, which is eclipsed every twenty-seven years by an invisible companion that blocks off half its light. Observations of the eclipses indicate that the obscuring body cannot be rounded, like a star, but must be a flattened disk. According to a recent analysis by mathematicians Michael Handbury and Iwan Williams of Queen Mary College, London, this secondary body is actually a protostar surrounded by a disk of dust in which planets may be forming. Thus Epsilon Aurigae helps provide support for the supposition that planets can arise in a binary system if the stars are far enough apart—in this case, about 3.5 billion kilometers, or midway between the distances of Uranus and Neptune from the Sun.

Are there any signs of planets that have already formed—solar systems like our own? Since planets around other stars cannot yet be seen directly, astronomers have to use indirect methods of detecting them. One such method is to search for the gravitational effect that an invisible star or planet has on the motion of its visible parent star.

Stars near the Sun show a systematic drift in position over the years, termed "proper motion," which results from their rotation around the Galaxy. Like cars on a giant racetrack, some stars move ahead of the Sun while others fall behind. A star's proper motion is imperceptible to the naked eye, but it can be detected on large-scale photographs taken with long-focus telescopes. A single star should move in a straight line; but if the star is double, or if it has planets, its motion will be unbalanced. It will therefore show a slight wavelike wobble in its path as it swings around in orbit with its invisible companion. The amount of the wobble and its frequency reveal the mass and orbital period of the invisible companion.

A detailed study of the motions of nearby stars, captured on photographic plates taken over the past forty years, has rewarded astronomers with the discovery of at least one possible planetary system, as well as many previously unknown low-mass stars. Peter van de Kamp of the Sproul Observatory, Swarthmore, Pennsylvania, began a regular photographic survey of the stars within 30 to 40 light-years of the Sun in 1937, using the observatory's 61-cm refractor.* In 1963

* The 100,000th photographic plate in the program was taken on March 1, 1974.

van de Kamp announced that the red-dwarf Barnard's star, the second closest star to the Sun, had a possible planetary system.

Barnard's star, lying 6 light-years away in the constellation Ophiuchus, has a mass 0.14 that of the Sun and gives out 0.00044 the Sun's light—too faint to be seen without a telescope. The wobble of Barnard's star on the Sproul photographs indicated the presence of a planet roughly 1.5 times the mass of Jupiter, orbiting every 24 years. Six years after his first announcement, van de Kamp raised the possibility that not one planet existed around Barnard's star but two, with masses similar to Jupiter and orbital periods of 26 and 12 years. Other astronomers suggested that even more planets might be present.

But it was later found that a spurious kink in the observed motion of Barnard's star had been caused in 1949 when the Sproul telescope's lens was placed in a new retaining cell made of cast iron, instead of the old aluminum one. This kink meant that previous interpretations of the movement of Barnard's star had to be modified, although the existence of a planetary system was not disproved.

The high-quality Sproul observations from 1950 onward have confirmed that the star's motion does waver. Using the accurate new data, Peter van de Kamp in 1975 published a revised analysis of the motion of Barnard's star in which he concluded that the existence of at least one planet "appears to be well established." Evidence for the existence of a second planet was, he said, "admittedly marginal," but helped satisfy the observations. He also noted that photographs of Barnard's star taken at the University of Pittsburgh's Allegheny Observatory and at the Van Vleck Observatory in Middletown, Connecticut, contain evidence for a wobble in the star's motion caused by at least one planet.

From data extending into 1976 Dr. van de Kamp has calculated the probable masses of the Barnard's star planets to be 0.9 and 0.4 that of Jupiter, with periods of 11.7 and 18.5 years, orbiting at distances of 410 million kilometers and 560 million kilometers. Planets like the Earth might also orbit Barnard's star, but their effects would be too small to be detected.

A second star with a possible planetary system or degenerate stellar companion found by van de Kamp is Epsilon Eridani, one of the nearby stars most like the Sun. Epsilon Eridani, 10.7 light-years away, is slightly cooler than the Sun, with three-quarters the Sun's mass and giving out one-third the light. Van de Kamp first proposed a massive planet-like companion for Epsilon Eridani in 1973. Using data into

1976, he finds that this companion has a period of 26 years and a distance from Epsilon Eridani of 1.2 billion kilometers. The estimated mass is at least six times that of Jupiter, large by solar system standards but below the minimum mass for a true star.

Other stars being tracked at Sproul for possible planetary systems are EV Lacertae (also cataloged as BD +43°4305) and BD +5°1668, both red dwarfs of about one-quarter the Sun's mass, 16 and 12 light-years away respectively. The red-dwarf star Lalande 21185, 8.1 light-years away in the constellation Ursa Major, was once credited with a planetary companion about ten times the mass of Jupiter, but follow-up studies have failed to confirm this. Other possible wobbles quoted in the past have turned out to be false alarms, although one star, 61 Cygni, remains controversial. It is a double star, the components having slightly over one-half the Sun's mass and orbiting each other every 720 years. The slightly larger and brighter star of the two, 61 Cygni A, was once suspected of having a planetary companion with a mass of about eight times that of Jupiter, but more recently it was crossed off the list.

Then, in 1977, two Soviet astronomers from Pulkovo Observatory, A. N. Deutsch and O. N. Orlova, reported the possible existence of three companions to 61 Cygni, with masses of 7, 6, and 11 times that of Jupiter and periods of 6, 7, and 12 years respectively. There is some indication, the astronomers report, that the two largest planets orbit one component of 61 Cygni, while the smallest planet orbits the other star. When their own data from Pulkovo was combined with more extensive data from Sproul, the calculated masses of the planets were halved.

The star has also been under survey by Murray Fletcher at the Dominion Astrophysical Observatory, Vancouver, B.C. He has used a sensitive spectroscopic method to search for the approximately 5-year orbital period of the planet around 61 Cygni A reported some years ago. In 1977, after five years of observations, he reported results which showed that if any planet exists around 61 Cygni A, it must have a mass smaller than 10 solar masses. It seems that the case with regard to 61 Cygni is too uncertain to draw any firm conclusions at present.

Photographic materials, plate-measuring techniques, and data analysis have all improved greatly since van de Kamp began his search in the 1930s. Astronomers can now track the tiny wavers in a star's motion with less effort and greater accuracy than ever before, thereby increasing the number of stars around which they might find planetary systems. A new search has in fact been started at Allegheny Observatory by a young astronomer, George Gatewood, who has assumed van de Kamp's mantle as a planet hunter.

The detectability of a planet depends on its mass, the mass of its parent star, and the star's distance from Earth. Gatewood estimates that planets the mass of Jupiter would be detectable with the Allegheny Observatory's 76-cm refractor around ten or more of the Sun's neighbor stars (see table). All these stars are red dwarfs, though planets larger than Jupiter could be detected around more massive stars. Answers are not expected for decades. Gatewood's search, like that of van de Kamp before him, may need to continue for a lifetime to be sure of the minute effects that are searched for. The slow pace of this research is dictated by the long times that planets take to go around their orbits; the largest planets of our solar system, Jupiter and Saturn, have orbital periods of 11.9 and 29.5 years respectively.

Stars Around Which Jupiter-like Planets
Are Detectable
(in Approximate Order of Detectability)

Star	*Distance* (light-years)	*Mass* (Suns)
Proxima Centauri	4.3	0.11
Barnard's star	6.0	0.14
Wolf 359	7.6	0.10
G 51-15	12.0	0.09
Ross 154	9.5	0.15
Ross 248	10.3	0.13
L 789-6	10.7	0.15
Ross 128	10.8	0.15
Groombridge 34B	11.6	0.15
Kruger 60B	12.9	0.13
G 158-27	14.6	0.10

One strong ray of hope for the future comes from the plans for large telescopes in space, which will enable astronomers to plot star positions with far greater accuracy than is currently possible through the dense, turbulent blanket of Earth's atmosphere. NASA's 2.4-meter space telescope, due for launching by the Space Shuttle in the 1980s, is expected to improve position-measuring accuracy ten times.

There has been an explosion of late in the number of feasible planet-spotting techniques being suggested. One method being introduced is a sensitive refinement of the technique for finding spectroscopic binary stars. The technique involves measuring slight changes in wavelength of light from a visible star as it swings around in orbit with its invisible companions. This wavelength change is known as the Doppler effect, and occurs in light from all moving bodies.

At the University of Arizona's Lunar and Planetary Laboratory, astronomer Kristopher Serkowski is monitoring the spectra of bright Sun-like stars for such tiny wavelength changes, using the laboratory's 155-cm reflector in the Santa Catalina Mountains near Tucson. He began by artificially polarizing different wavelengths of light from the star to assist in measuring any slight wavelength changes, but in 1976 switched to a device known as a Fabry-Perot interferometer, which transmits only certain very narrow wavelength intervals. Any slight wavelength changes are registered by a detector.

Serkowski hopes to measure star motions as small as 5 meters a second, which would be sufficient to reveal the existence of planets the size of Jupiter around his target stars. In our own solar system, the Sun is slightly unbalanced by the presence of Jupiter, which is the only planet that matters, since it is two and a half times the mass of all the other planets put together. The Sun and Jupiter act like the two halves of a very uneven double star. Although it is true for most practical purposes to say that Jupiter orbits the Sun, when dealing with precision astronomical measurements one must also take into account the fact that the Sun moves around the center of gravity of the Sun-Jupiter system at about 12.7 m/sec. Therefore a distant observer would see a cyclical change in the wavelength of light from the Sun of the type that Serkowski is seeking. This is in fact another, more sensitive way of finding the same star wobbles as sought by photography. Unlike the photographic technique, it is not restricted to the nearest stars.

For his study, Serkowski has selected twenty-five stars similar to the Sun. All are brighter than magnitude 6, appear nearly overhead at

Tucson, and have no known stellar companions. In addition, he is extending the work of Abt and Levy by observing the fifty-two stars for which they found no detectable companion. Each star needs to be observed on at least four nights a year to obtain the desired accuracy of 5 m/sec in its motion. Serkowski estimates that the program will take a decade to complete.

A similar project is planned by Dr. George Isaak of Birmingham University, who has developed a system that accurately compares specific wavelengths in a star's spectrum with a wavelength standard on Earth. In trials of this system, Isaak and his team have found what they claim to be regular oscillations of about 10 kilometers in the outer layers of the Sun.

Another sophisticated approach has been suggested by Harold McAlister of Kitt Peak Observatory. He has noted that it should be possible to detect the presence of any planets in binary star systems by the technique known as speckle interferometry, which unscrambles the blurring caused by Earth's atmosphere to give the accurate relative positions and orientations of the members of close double stars. Planets themselves could not be seen directly with this technique, but speckle interferometry promises to give star positions in binaries at least fifteen times as accurately as with direct photography, thereby allowing astronomers to search for slight wobbles in the paths of the components of double stars caused by the presence of planets.

In 1976 a team of astronomers and engineers met for a summer study on planetary detection at NASA's Ames Research Center in California. During the study, called Project Orion, they came up with a proposal for a so-called imaging interferometer, a twin telescope whose two views of a star field are combined to give more accurate positions of stars than are possible with a single telescope alone. Such a twin telescope, either on Earth or in space, could detect the wobbles in star motions that would reveal the existence of Earth-sized planets around about 400 stars out to a distance of 32 light-years, or Jupiter-sized planets around most stars out to 150 light-years, which is vastly better than the single-telescope astrometric studies being made by van de Kamp and Gatewood.

But what are the chances of seeing planets of other stars directly? The problem is that the faint light from planets is swamped by the light, often 1 billion times brighter, of their parent stars. Even space telescopes, such as the 2.4-meter reflector due to be launched by the Shuttle, will not be able to resolve planets on its own. But Charles KenKnight of the Lunar and Planetary Laboratory at the University of Arizona has pointed out that special optical systems can be devised to cut down the amount of light from parent stars, thus revealing images of any bright accompanying planets amid the glare. He believes that a 2-meter telescope in Earth orbit equipped with suitable optics could see planets on the order of Jupiter or even Venus (because of its brightness) around nearby stars.

As the results from these searches come in over the next decade, astronomers will for the first time have observational evidence to back up their theoretical speculation that planets are common in space. For the moment, though, only one star is known with certainty to have planets—the Sun.

Ironically, although stars provide the heat and light to nourish any life around them, they also drive that life away when they die. About 5 billion years from now, the Sun will begin to heat up at the center. It will slowly swell to a red-giant star, 100 times its present diameter, engulfing the planets Mercury, Venus, and perhaps even Earth. Then it will puff off its outer layers into space, leaving its hot core as a slowly cooling white dwarf.

Long before that happens, of course, all remaining life on Earth will have been incinerated. Once the Sun starts to heat up it will cause widespread changes in Earth's climate, a melting of the polar caps, and consequent flooding of lowland areas. Then the seas will evaporate, leaving the Earth a parched and bare cinder. Any distant descendants of ours will have left Earth in search of another home around a younger star. Other civilizations, if any exist, may already have been through this enforced emigration from their home star.

Once a star starts to die, the existence of planets around other suns becomes more than a matter of academic curiosity. It is a matter of life or death.

REFERENCES

Abt, H. A., *Scientific American*, April 1977, p. 96.
——— and S. Levy, *Astrophysical Journal Supplement*, Vol. 30 (1976), p. 273.
Deutsch, A. N., and O. N. Orlova, *Soviet Astronomy*, Vol. 21 (1977), p. 182.
Fletcher, M., *Science Dimension*, Vol. 9 (1977), p. 18.
Gahm, G. F., et al., *Astronomy and Astrophysics*, Vol. 33 (1974), p. 399.
Gatewood, G., *Icarus*, Vol. 27 (1977), p. 1.
Handbury, M., and I. P. Williams, *Astrophysics and Space Science*, Vol. 45 (1976), p. 439.
Hartmann, W. K., *Astronomy*, September 1977, p. 6.
Heppenheimer, T. A., *Astronomy and Astrophysics*, in press, 1978.
Herbig, G. H., *Mercury*, Vol. 5 (1976), p. 2.
Hohlfeld, R., and Y. Terzian, *Icarus*, Vol. 30 (1977), p. 598.
Isaacman, R., and C. Sagan, *Icarus*, Vol. 31 (1977), p. 510.
KenKnight, C. E., *Icarus*, Vol. 30 (1977), p. 422.
McAlister, H. A., *Icarus*, Vol. 30 (1977), p. 789.
Ney, E. P., *Sky & Telescope*, Vol. 49 (1975), p. 21.
Project Orion, *Spaceflight*, Vol. 19 (1977), p. 90.
Serkowski, K., *Icarus*, Vol. 27 (1976), p. 13.
Thompson, R. I., et al., *Astrophysical Journal*, Vol. 218 (1977), p. 170.
van de Kamp, P., *Annual Review of Astronomy and Astrophysics*, 1975, p. 295.
——— *Astronomical Journal*, Vol. 79 (1974), p. 491.
——— *Astronomical Journal*, Vol. 80 (1975), p. 658.
——— *Vistas in Astronomy*, Vol. 20 (1977), p. 501.

3

Life in the Solar System

The nine planets and associated debris of our solar system provide a vast range of conditions, from searing heat to near-absolute cold, from total vacuum to dense gas, from zero gravity to many times Earth gravity, and from arid sterility to a seething soup of organic chemicals. If, as seems probable, there are other planetary systems in space, there is likely to be something of these conditions in every one of them. By studying our solar system, we should gain clues to the possible niches in which life may form in space. This study will illuminate the third unknown in the Drake equation—the number of habitable planets in each solar system.

Mercury, the closest known planet to the Sun (there may be some asteroids closer still), is probably typical of the smaller planets in space, at least on its outside. Internally, it is remarkable in having a giant iron core, comprising about four-fifths its total diameter; the core is made from the heavy elements that remained closest to the Sun in the cloud from which the planets formed.

Mercury's rocky surface, first shown in detail in 1974 by the Mariner 10 spacecraft, has been badly battered by impacts from debris left over after the formation of the solar system. Mercury's craters, basins, and lava plains are very reminiscent of the Moon—not surprisingly so, for Mercury's diameter of 4,880 kilometers is only 50 percent larger than that of the Moon; some of the satellites of the outer planets are larger than Mercury. The Moon and Mercury do not have sufficient gravity to hold an atmosphere that would shield them from meteorites.

Mercury turns on its axis once every 59 days, two-thirds of the time it takes to orbit the Sun. The remarkable consequence of this is that from one Mercurian noon to the next takes 176 days; during this time, the planet has orbited the Sun twice and spun three times on its axis. Mercury's rotation has been braked by the gravitational pull of the Sun.

Standing among the lunar-like craters on the day side of Mercury, one would see the Sun two and a half times larger than it appears from Earth. Without the filtering effect of an atmospheric blanket, one would be bathed in deadly levels of short-wavelength radiation as the Sun heated surface rocks to 400°C or more at maximum, hot enough to melt tin and lead. But during the three-month Mercurian night, the temperature of the surface rocks would drop to −170°C.

Mercury is simply too close to the Sun for life to exist on its surface. If the Sun were slightly cooler, a planet with an atmosphere at Mercury's distance would be at the right temperature for life. But around the smallest and coolest stars of all, red dwarfs like Barnard's star, a planet would have to be even closer than Mercury is to the Sun to be kept warm enough for life—so close that its axial spin would be locked by gravity so that it kept one face forever turned toward the star, as the Moon keeps one face turned to the Earth. Life on a planet with permanent day in one hemisphere and eternal night in the other would be a little monotonous, but not necessarily impossible.

How near to a star or how far from it must a planet be to support life? Each star is surrounded by a habitably warm zone known as its ecosphere in which, as around a bonfire, temperatures are neither uncomfortably hot nor uncomfortably cold. The size of a star's life zone depends on the star's temperature: the coolest stars, like faintly glimmering bonfires, warm only a small area, whereas the hottest stars spread heat and light over a much larger area. Our Earth, not surprisingly, is at the very center of the Sun's life zone. Venus, the second planet from the Sun, is just at the inner edge of the habitable zone, and it is instructive to compare the differences between Venus and Earth to see how finely the scales are balanced between an uninhabitable planet and one that teems with life.

On the face of it, Venus and Earth appear very similar. Venus is only slightly smaller than Earth (12,104 km diameter, as against 12,756 km), and they come closer together (within 41 million km) than any other two planets. Both Venus and Earth have copious atmospheres. A long-standing nickname for Venus was "Earth's Twin." But the results of space-probe research have shown what a devilish misnomer this is. For Venus has more in common with Hell than with our own Earthly paradise.

Venus appears brighter in the sky than any other planet because sunlight is reflected strongly from the dense clouds that obscure the planet's surface from the prying eyes of astronomers. Only radar can see through the clouds, to produce crude maps of the surface which show Venus to be peppered with craters, cut by canyons, layered with lava flows, and crested by mountains that look suspiciously similar to the giant volcanoes of Mars. Apparently Venus has had an active geological history.

One distinct curiosity about the planet is that it rotates on its axis every 243 days from east to west, instead of the usual west to east, longer than the 225 days it takes to orbit the Sun. From Venus, therefore, the Sun would appear to rise in the west and set in the east, taking 117 days to go once around the sky (half Venus's year). The rogue back-to-front rotation of Venus is a major puzzle, but it may somehow be connected with the combined gravitational pulls of the Sun and Earth.

The clouds themselves are now believed to be composed of a strong solution of sulfuric acid, in about 80 percent concentration. The cloud tops visible in telescopes are at a height of 60 km above the planet, and according to spacecraft measurements extend down to 35 km from the surface. Below that is a thick atmosphere of almost pure carbon dioxide, which reaches a pressure of 90 Earth atmospheres at the surface—equivalent to the pressure half a mile under the ocean on Earth, sufficient to crush a submarine like tinfoil.

Not surprisingly, the first Soviet probes to parachute into Venus's atmosphere were destroyed before they ever reached the ground. Only in 1970 were the first signals received from the surface of Venus, transmitted by the Venus 7 probe. Subsequent probes have sent back data indicating that the rocks on Venus are similar to granite and basalt on Earth, and have taken photographs of the rocky surface which show that as much light penetrates the clouds as does on a cloudy summer day on Earth.

These probes have confirmed previous findings that the temperature at the surface of Venus, in both day and night hemispheres, is a furnace-like 475°C. A luckless astronaut who crash-landed on Venus would thus be simultaneously crushed, roasted, and suffocated.

Two planets, similar in size, in the same part of the solar system, and apparently of similar composition—why should Earth and Venus today have such startlingly different conditions? Venus, by virtue of its proximity to the Sun, receives nearly twice as much solar radiation as Earth; but 80 percent of this is reflected by the clouds, so the surface temperature should be comfortably warm. Why, then, is Venus so roastingly hot? And where has its water gone?

One point of similarity between Venus and Earth, not obvious at first glance, is that both planets have comparable amounts of carbon dioxide; on Earth, though, most of it is locked up in limestone sediments, a result both of natural chemical processes and of the action of sea creatures in the past. Carbon dioxide traps heat very efficiently, in a process known as the greenhouse effect (indeed, one of the fears on Earth is that the extra carbon dioxide being released into the atmosphere by the burning of fossil fuels may cause drastic changes in our climate). Sunlight can warm Venus in the normal way, but the presence of so much carbon dioxide in the atmosphere prevents the heat from escaping again, thus building up the planet's high temperature.

Both Venus and Earth are believed to have obtained their present atmospheres from gases released from their interiors, particularly through the mouths of volcanoes. The volcanoes broke through the crust as the interiors of the planets heated up from the decay of radioactive atoms. Bodies such as Mercury and the Moon would have been too small to heat up sufficiently, which is why they have no large volcanoes.

This degassing through volcanoes would presumably have produced similar gases on both planets—mostly water vapor, carbon dioxide, and nitrogen. But already the evolution of the two planets' atmospheres was being altered by their different temperatures. Venus, being closer to the Sun, would have been hotter than Earth to start with, and this higher surface temperature would have affected the chemical reaction that binds up carbon dioxide in the planet's crust. While chemical reactions on Earth removed carbon dioxide from the atmosphere and kept the planet down to a reasonable temperature for life to form, on Venus this process would not have occurred. The carbon dioxide would thus have built up in the Venus atmosphere, making the planet still hotter, and leading to the oven-like temperatures of today.

What of the water on Venus? The main product of volcanoes on Earth is water vapor, which has condensed to fill our oceans. But water on Venus today is almost nonexistent. The reason seems to be quite straightforward. Since Venus is so hot, its water would have remained as vapor in its atmosphere, where it would be broken up by sunlight. This process, called photodissociation, splits water into its component hydrogen and oxygen atoms. The hydrogen, being lightweight, would be lost into space, while the oxygen could both combine in chemical reactions with the crust and help to make the sulfuric acid clouds. Water vapor in the atmosphere would have worked with the carbon dioxide to block the loss of heat from Venus, thus leading to a runaway greenhouse effect that quickly produced the hot, arid, sterile planet we see today.

It is possible that in the past Venus had oceans. Theories of stellar evolution indicate that the Sun would have been slightly cooler thousands of millions of years ago, so that the greenhouse effect on Venus may not then have got under way. Venus, in fact, may once have been quite Earth-like, with volcanoes belching gases into the sky and cooling streams running over the faces of continents to fill the oceans. In those oceans, fed by a rain of organic molecules from the planet's chemically rich atmosphere, life could have begun to evolve at the same time as it did on Earth, nearly 4 billion years ago. But eventually, after perhaps 3 billion years, the warming Sun would have snuffed out that life as the planet's climate changed catastrophically. The fossils of dead sea creatures may lie in the rocks of the deep basins on Venus, mute testimony to a planet that, a few million kilometers farther from the Sun, could have been as fertile as Earth.

One day, Earth will become like Venus. As the Sun swells into a red-giant star at the end of its life, in billions of years' time, it will begin to heat up Earth. As Earth gets hotter, the carbon dioxide will be liberated from the rocks in which it is chemically trapped. Water will evaporate from the seas, be broken up by sunlight, and be lost.

The runaway greenhouse effect will ensue in the predominantly carbon dioxide atmosphere, and Earth will become a burning Hell.

The strong sulfuric acid clouds of Venus and the lack of water (no more than one part per thousand, possibly one hundred times less) seem to rule out the suggestion made some years ago that Venus could be made habitable by seeding its atmosphere with algae that would break down the carbon dioxide to release oxygen. The situation might be saved if we could arrange to import more water—perhaps by diverting an icy comet into the clouds of Venus.

Liquid water is the one factor which makes Earth the most favorable planet in the solar system for life. Water itself is not lacking in the solar system; its simple combination of two hydrogen atoms with one of oxygen is plentiful, particularly on the giant outer planets. But there it is either frozen as ice or in droplets as water vapor. On Earth it is abundant as a liquid because our planet lies in the middle of the Sun's life zone—that celestial green belt in which we expect to find a star's life-bearing planets.

Water is the liquid of life. It acts as a solvent in which chemicals can come together to react; as far as we know, life originated in or around the oceans of the early Earth. Water is also the major component of living things, about 90 percent in the case of many vegetables (think of instant potatoes or dehydrated soups), and two-thirds in the case of human beings. Without water, organisms cannot live.

According to the theories of the origin of life, the basic organic chemicals which comprise proteins and nucleic acids were made easily in the atmosphere of the primitive Earth; many of those organic chemicals have been produced in experiments that simulate conditions 4 billion years ago. And those component chemicals are the same in the proteins and nucleic acids of every form of life on Earth—only their arrangement is different, like different shapes made from the same set of building blocks. Such similarity at the microscopic level gives a truly breathtaking glimpse of the basic unity of all life on this planet, from microbes to man.

Proteins are the structural material of life, whereas the nucleic acids, DNA and RNA, guide reproduction and growth (the genetic code). Cells are, at their most basic, a shell of protein enclosing nucleic acids. How the first proteins and nucleic acids were built, and how they came together in cells to form the first living organisms, are unsolved problems that remain at the forefront of biochemical research.

Chemical fragments of proteins and nucleic acids seem to have helped in each other's development, a two-way process which eventually built up today's full-sized proteins and nucleic acids. Proteins and nucleic acids still bear the imprint of their intimate evolutionary relationship, as shown by the model-building experiments of American biologists Charles Carter and Joseph Kraut, who found that twisted chains of protein wrap neatly around the famous double-helix spirals of nucleic acid.

Leslie Orgel and his colleague André Brack of the University of California's Salk Institute at San Diego concluded that the first protein would most likely have been made of alternating chains of the amino-acid chemicals glycine and alanine, which are the most abundant amino acids formed in simulations of conditions on the primitive Earth. The genetic code for this protein would require a chain of alternating nucleic acids, as observed in the nucleic acids of today. Thus the genetic code would have originated by interactions between the first simple proteins and the first simple chains of nucleic acids.

The vast complexity of the origin of life has been amply depicted by organic chemist Dr. Peter Molton in an outline of eighteen major steps to the formation of a cell, involving the synthesis and amalgamation of chemical units that developed separately in the pre-biotic soup of the Earth's young oceans. Each of those first chemical units was probably the barest arrangement capable of doing the job; they were improved by later additions, producing the highly complex molecules we see today. Some cells incorporated a unit that allowed them to use carbon dioxide as food—the reaction known as photosynthesis. Thus came the division of life into plant cells (the photosynthesizers) and animal cells (those that preyed on the plant cells, and on each other). Eventually, cooperation between cells produced the astounding multicellular organisms that dominate Earth.

Fred Hoyle and Chandra Wickramasinghe in their "black cloud" paper of 1977 joined the ranks of those who propose that Earth may have been seeded by life from outer space. Astronomers know that the dark clouds in space contain microscopic graphite and silicate dust, and about forty molecules of varying complexity. Hoyle and Wickramasinghe quoted evidence from the absorption of light at ultraviolet wavelengths by dust in space that even more complex molecules, including the molecules of life, can be built up on the surfaces of the dust grains in clouds such as the Orion nebula. Rocky meteorites, such as that which landed at Murchison, Australia, in 1969 have been found to contain quite complex organic molecules, including amino acids, apparently formed in space under the conditions that Hoyle and Wickramasinghe describe.

Later in 1977, Hoyle and Wickramasinghe noted that the absorption of light by dust clouds in space at certain infrared wavelengths reveals the apparent existence of cellulose in space, formed from stable rings of the organic molecule formaldehyde (H_2CO), which is already known to be abundant. Cellulose, of course, is the structural material of plant cell walls.

Hoyle and Wickramasinghe propose that the clumping and breaking up of dust grains in the interstellar clouds could eventually produce simple forms of living organisms, trapped and preserved inside grain clumps and feeding off the nutrient of the surrounding gas. These organisms would divide and grow like any normal biological system as the grain clumps split and recombined, except that once a grain clump and its kernel of organic material was isolated from the surrounding cloud it would become biologically inert, rather like a spore. Such a spore, if it were contained in a meteorite, would remain in a state of suspended animation until released in a suitable planetary atmosphere, when it would spring back to life and begin to evolve in its new environment.

More recently, Hoyle and Wickramasinghe have argued that comets are favored sites for the origin of life, and that not only may life have been started by an impact of a comet with Earth, but that encounters with micrometeorites from comets in modern times are responsible for delivering periodic waves of disease to Earth. On this basis, life would be expected to be frequent throughout the Galaxy, and also to be built on biochemical lines similar to our own.

Most biologists remain skeptical of the view that Earth was seeded with life from space; they feel that conditions were far more conducive for life to have evolved on Earth in the first place. Also, no "living" spores have been detected in meteorites, and the failure to detect life on Mars seems to argue against the proposal.

Despite these advances in unraveling many of the steps to life, no scientist is anywhere near being able to create life in the laboratory. Until that can be accomplished—and it may eventually happen—the origin of life remains a mystery. And while it remains so, the corresponding factor in the Drake equation is totally uncertain.

It has traditionally been assumed that life might arise on one planet in ten with suitable conditions. But this is nothing more than an assumption—an expression of faith, some might call it. This is why the hunt for traces of life on another planet assumes such overriding importance. Evidence that life has arisen spontaneously on another body in space would place our existence in perspective more sharply than any other discovery we could make. It would, in the words of physicist Philip Morrison, turn the origin of life from a miracle into a statistic. One point of optimism is that traces of the earliest, single-cell organisms on Earth extend back 3.5 billion years or possibly longer, thus indicating that life, however caused, did not take long to get going.

Two other factors in the Drake equation—the probability of intelligence and the frequency of technological civilizations—are as uncertain as the origin of life. Intelligence took much longer to develop on this planet than life itself. Only during the past few million years have beings of the *Homo* lineage been present on Earth; latest discoveries of fossil man place the split of true man from the ape-men at 3 to 4 million years ago. By contrast, our development from the use of stone tools to the level of high technology has been the fastest stage of all in our evolution. Some biologists believe that high intelligence is such an advantage that it is almost certain to develop, given time. And the inquisitive, dexterous nature of highly intelligent beings is bound to make them into a technological civilization that starts to question and examine its surroundings, probing even for evidence of other beings elsewhere in space. But further discussion of these matters is for anthropologists and sociologists, and we must return to the solar system.

Throughout this century the planet Mars, next past the Earth, has seemed the most likely place to find evidence of other life in space. In the popular vision, constructed by astronomers such as Percival Lowell at the turn of the century and eagerly supported by science-fiction writers since, Mars was a dying world whose inhabitants were fighting a losing battle against the planet's parched deserts. Their technology had produced artifacts visible clear across the multi-million-kilometer gap to Earth: irrigation canals for channeling melt water from the polar caps to their crops near the equator. Lowell and his followers drew an intricate network of canals, interspersed with dark patches they termed oases, and charted what appeared to be vast areas of crops which varied seasonally in extent and coloration.

The large dark markings were real enough; moderate-sized amateur telescopes will show the largest of them, and they have been captured clearly on photographs. But what of the canals? They first came to prominence after the close approach between Mars and Earth in 1877 (the same occasion, incidentally, on which the two moons of Mars were discovered). The Italian astronomer Giovanni Schiaparelli drew a series of linear markings on the planet which he termed channels. Schiaparelli always kept an open mind about their true nature, but for the American astronomer Percival Lowell they were clear signs of an advanced technological civilization on the planet.

Lowell, a member of a famous New England family, built his own observatory in the dry, clear air of northern Arizona to study Mars. During the years following the observatory's opening in 1894, Lowell and his collaborator Earl C. Slipher charted an increasingly complex network of canals. These findings, coupled with Lowell's romantic visions of the decaying civilization on Mars, fired the public's imagination—despite the fact that most other observers failed to see the canals, and denied their existence.*

After Lowell's death in 1916 the debate rumbled on, with most astronomers regarding Lowell's claims with increasing skepticism. The dispute was finally laid to rest in 1965 with the arrival of the Mariner 4 spacecraft, which radioed back to Earth a string of photographs as it sped past Mars. The photographs showed a cratered surface that looked more like the Moon than the Earth. Of the canals—or of life—there was no sign. The Mariner 6 and 7 craft, which photographed more of Mars in 1969, were no more encouraging, indicating that Mars was a dry, frigid, and near-airless world.

But a full photographic survey had to wait until 1971–72, when the Mariner 9 reconnaissance probe went into orbit around Mars. As expected, the large dark and light patches noted by Earth-based observers

* The Lowell Observatory at Flagstaff, Arizona, is today one of the foremost centers of planetary research. The astronomers there now seem a little embarrassed about the observatory's sensational early history.

were found to correlate with areas of darker and lighter rock or dust on the planet. But where were the fine, straight lines drawn by the canal observers? Cornell University astronomers Carl Sagan and Paul Fox overlaid Earl C. Slipher's definitive map of the canals onto a mosaic of Mariner 9 pictures. In only a few cases did the "canals" correlate with real surface features, such as a chain of craters or a dark surface streak, and in many instances there were topographic features or markings which had not been recorded by Slipher despite the fact that they were much wider than many of the supposed canals which he drew. Basically, therefore, it seems that most of the canals are unexplained. Sagan and Fox characterized the canals as monuments to the imprecision of human observers under difficult conditions.

Any hopes of finding patches of vegetation on the surface of Mars were finally dashed by the Mariner 9 pictures. The changes in shape and intensity of the dark markings are caused by dust being blown around by seasonal winds, not growing vegetation.

But the space probes did not completely rule out the possibility of life on Mars. In fact, they discovered one critical factor which made it more likely: Mars has volcanoes, and spectacular ones too. The main volcanic mountain on Mars, Mount Olympus, is the largest known volcano in the solar system, dwarfing even the volcanic Hawaiian islands on Earth. Volcanoes on Mars opened the prospect of water on the planet, and Mariner 9 photographed numerous examples of what looked suspiciously like dried-up rivers, which presumably flowed during more clement times in the planet's history.

The present atmosphere of Mars is scanty. It is made up mostly of carbon dioxide gas, as thin as the Earth's air at a height of 32 kilometers. Consequently, temperatures on Mars are well below freezing, but the Mariner 9 findings held out the hope that Mars may have had a denser atmosphere in the past, particularly at the time when the volcanoes were active. With a denser atmosphere, Mars would also have been warmer. Had Mars been a slightly larger planet—it is half the diameter of Earth—it would have become hotter inside and degassed more, building up a substantial atmosphere that could have kept it comfortably warm to support life. On this view, Earth at the distance of Mars could still be a habitable planet.

Even so, Earth-type organisms could just about cling to survival under present-day Martian conditions. Optimistic biologists believe that life might once have arisen on the planet, and that traces of it might still be found in the Martian soil. The two American Viking probes were designed to find out. Their results are described in the next chapter.

Beyond Mars, the solar system grows colder. This is the realm of the giant planets—and none is larger than Jupiter, king of our solar system, containing over twice as much mass as all the other planets wrapped up together. In contrast to the four small, solid bodies of the inner solar system, the four giant planets of the outer reaches are mostly made of gas, similar in composition to the Sun. Jupiter, in particular, is like a star that failed. It still releases a pale warmth of its own, possibly caused by a continual overall shrinkage of about 1 millimeter a year.

When we look at the outer planets, all we see is a cloudy atmosphere. The thick cloak of gases around each of these planets is a remnant of the cloud around the Sun from which the solar system first formed. Similar primitive atmospheres may also have surrounded the inner planets shortly after their birth, later being blown away by solar radiation. By contrast, the giant planets, being larger and cooler, retained the gases of their primitive atmospheres. The main components of Jupiter, as of all the outer planets, are hydrogen and helium, the two lightest and most abundant substances in the Universe. Helium is chemically unreactive, but small percentages of other substances, in a body the size of Jupiter, provide plenty of molecules for chemical reactions. The most abundant reactive materials on Jupiter are hydrogen, methane, ammonia, and water—those same constituents from which life formed in the Earth's early atmosphere. Jupiter is thus a kind of space fossil. When we look at it we are viewing conditions on our Earth 4 billion years ago.

Simulations of Jupiter's atmosphere, like the simulations of primitive Earth, have shown that organic molecules are plentifully formed from the gases by the action of ultraviolet light, electric sparks, and shock waves. The atmosphere of Jupiter provides 1,000 times the volume for organic synthesis than the atmosphere of Earth. Many of the resulting molecules are highly colored, from red (azobenzene, the starting point of many organic dyes) to blue (azulene), which can help explain the welter of transitory colors seen in the clouds of Jupiter. Evidently organic synthesis in Jupiter's atmosphere continues today, and traces of such molecules should be detected by the infrared spectrometers being carried by the Voyager 1 and 2 probes, due to reach the planet in March and July 1979.

But the presence of organic molecules, formed by simple chemistry, does not of itself imply the existence of life. The white cloud tops of Jupiter are made of frozen ammonia at a temperature of −120°C; this is not a likely habitat for life. Slightly below the clouds, however, temperatures may be warm enough for liquid ammonia droplets, which raises the question of alternative biochemistries to those we know on Earth. A plausible arrangement for life is to use liquid ammonia as a solvent instead of water, and it may be that the atmospheres of the outer planets provide the ideal conditions for a liquid-ammonia life form.

The ammonia clouds form Jupiter's turbulent weather layer, responsible for the continual changes in the planet's appearance visible in telescopes. Below this is a much more stable region where temperatures lie between the freezing and boiling points of water, at pressures that increase with depth from five to ten times the Earth's atmospheric pressure. Here, about 80 kilometers below the visible cloud tops, clouds of water vapor form, making this the most promising locale for Jovian life. Lower still, temperatures and pressures increase rapidly until the hydrogen of which the planet is predominantly made is compressed first into a liquid, then into a superdense state with the properties of a metal. Jupiter has no solid surface as we know it on Earth; any biology must therefore be restricted to its clouds.

Carl Sagan has joined with Cornell University colleague E. E. Salpeter, an astrophysicist, to examine the possible ecologies of Jupiter's atmosphere, and they conclude that there is an "abundant biota" in the Jovian clouds. They compare the warm, wet cloud layers with the seas of Earth, which have simple photosynthetic plankton at the top level, fish at lower levels which feed off these creatures, and marine predators which hunt the fish. The hypothetical Jovian equivalents of these organisms are respectively termed sinkers, floaters, and hunters.

The sinkers will drift downward from just below the level of the visible clouds into the deeper atmosphere of Jupiter, eventually being burned up by the planet's internal heat. Before then, they must have reproduced in order to maintain a steady-state population.

As the sinkers drift downward, they provide food that is mopped up by the floaters. Organisms that can propel themselves around to coalesce with other bodies in the clouds of Jupiter are the hunters. The approach and coalescence of two organisms serves to exchange genetic material—the same function as mating. Some highly evolved Jovian organisms may include aspects of all three life styles in their life cycles.

The creatures that Sagan and Salpeter envisage are gas bags that move by pumping out helium, retaining the hydrogen for balloon-like buoyancy. Hunters in the clouds of Jupiter could grow to be many kilometers across, within the resolution of the cameras carried by the Voyager 1 and 2 probes.

If Sagan and Salpeter are right, we may already have visible evidence of life in Jupiter's clouds; they say the clue lies in the red coloration so frequently seen on the planet. Their estimates show that the direct action of sunlight does not seem to provide adequate supplies of red-colored molecules at the right levels in the atmosphere, below the white ammonia clouds. Instead, they claim that sufficient red coloration results from biological activity by Jovian organisms capable of photosynthesis, like simple algae on Earth. In other words, Jovian organisms are themselves red, like certain colored plankton on Earth that give the Red Sea its name. The great red spot, an updraft of heated air with prominent coloration, may thus be particularly favored for life.*

Sagan and Salpeter do not speculate on how life may have arisen on Jupiter in the first place. One possibility is that, on a planet with no solid surface, organic molecules may join together into forms of life on the surfaces of tiny particles, both liquid and solid, floating in the atmosphere. While the basis of Jovian life would probably be carbon, which is contained in such substances as methane, ethane, and acetylene in the planet's atmosphere, other details of Jovian biochemistry are completely uncertain. As Sagan and Salpeter say, it is possible there exists a variety of different paths to the origin of life, some of which favor the Earth while others favor Jupiter.

An entry probe into Jupiter's atmosphere carrying sensitive instruments could detect the level of organic molecules predicted by the biological hypothesis of Sagan and Salpeter. The current Voyager

* Conventional astronomy attributes the spot's color to red phosphorus.

spacecraft will not enter the atmosphere of Jupiter; that will be left for a planned orbiter/entry mission in 1982. Voyagers 1 and 2 will photograph Jupiter as they fly past before proceeding to the ringed planet Saturn, which they will reach in November 1980 and August 1981.

Saturn, like Jupiter, is mostly made of hydrogen and helium, and also seems to have an internal heat source. Saturn has dusky-brown and yellow bands between white belts of ammonia clouds, but the markings on Saturn are less prominent and less variable than the turbulent clouds of Jupiter. Also present in the atmosphere is a small percentage of methane, from which the action of sunlight builds up traces of other hydrocarbons such as ethane, ethylene, and acetylene. More complex organic molecules are also likely on Saturn. The similarity between Saturn's and Jupiter's atmospheres leads Sagan and Salpeter to propose possible airborne organisms in the clouds of Saturn, although with less certainty.

Some of the satellites of the giant planets are possible sites for synthesis of organic molecules, notably Saturn's largest satellite, Titan, which has an appreciable atmosphere. But if there is no life on the far more Earth-like surface of Mars, there is unlikely to be life of any kind on these small bodies.

Our knowledge of conditions on the planets beyond Saturn is woefully slight, but what we do know is not promising for life. The outer gas giants, Uranus and Neptune, show virtually no features in their greenish methane atmospheres, thereby indicating an almost total absence of the turbulent mixing that makes Jupiter's clouds so chemically complex.

There may be seas of liquid methane deeper under these clouds, where low-temperature life might survive, but sunlight at these distances is so weak that the synthesis of complex organic compounds or the origin of photosynthetic organisms must be considered highly unlikely. There is a hint that Neptune, like Jupiter, produces internal heat, which may improve matters slightly. But overall these remote, frigid bodies seem forbidding to life.

A group of American scientists, led by biologist Lynn Margulis of Boston University, reported in 1977 that the probability of a terrestrial organism being able to survive in the atmosphere of Uranus or Neptune is nil. There seem formidable barriers to the growth of Earth organisms even in the more hospitable atmosphere of Titan. Therefore the chances of life having arisen under these conditions must also be zero.

If the two Voyager probes to Jupiter and Saturn are successful, the second will be sent on to examine Uranus, arriving there by January 1986. We should then learn more about the prospect of life in the outer reaches of the solar system.

REFERENCES

Carter, C., and J. Kraut, *New Scientist*, Vol. 65 (1975), p. 784.
Hoyle, F., and N. C. Wickramasinghe, *Nature*, Vol. 268 (1977), p. 610.
——— *New Scientist*, Vol. 76 (1977), p. 402.
Margulis, L., et al., *Icarus*, Vol. 30 (1977), p. 793.
Molton, P. M., *Journal of the British Interplanetary Society*, 1978 (in press).
Orgel, L., and A. Brack, *Nature*, Vol. 256 (1975), p. 383.
Ponnamperuma, C., *Icarus*, Vol. 29 (1976), p. 321.
Sagan, C., *Space Science Reviews*, Vol. 11 (1971), p. 827.
——— and P. Fox, *Icarus*, Vol. 25 (1975), p. 602.
——— and E. E. Salpeter, *Astrophysical Journal Supplement*, Vol. 32 (1976), p. 737.
Wickramasinghe, N. C., *New Scientist*, Vol. 74 (1977), p. 119.

4

What Viking Found on Mars

The Viking 1 space probe landed on Mars on July 20, 1976, in the most difficult technical achievement in the history of the space age. Its landing site was the flat lowland known as Chryse, the Plain of Gold. Its mission, and that of its sister craft Viking 2, which landed on September 3, was to answer the question: Is there life on Mars?

Each Viking lander carried a tiny biological laboratory to analyze Martian soil collected by the craft's remote-controlled sampling arm. By incubating the soil in three different ways and seeing if anything grew, the biologists hoped they would find whether or not there was any life on the Martian surface. One experiment, called the gas-exchange experiment, assumed that Martian organisms, like terrestrial organisms, would respond to the addition of water and a rich mixture of nutrients. It simply attempted to detect changes in the gases above a sample of Martian soil which was fed this nutrient solution, colloquially termed "chicken soup"; the proportions of various gases would be expected to change as any Martian bugs breathed, ate, grew, and reproduced.

Two other experiments looked for the release or uptake of carbon by Martian microorganisms. One of these, the labeled-release experiment, fed a Martian soil sample with a nutrient solution containing radioactively labeled carbon; any carbon dioxide or other carbon-containing gas released by the sample would be detected by radioactive counters. The third experiment, the pyrolitic-release experiment, looked for evidence of photosynthesis in the soil, as would be expected if plantlike organisms existed there. In photosynthesis, carbon is taken up from the atmosphere, which on Mars consists mostly of carbon dioxide (a trace of carbon monoxide also exists). In the pyrolitic-release experiment, soil samples were incubated under an artificial Martian atmosphere of carbon dioxide and carbon monoxide which had been labeled with radioactive carbon; the soil was then heated (pyrolyzed) to drive off any carbon that may have been taken up by microorganisms, and the resulting gases were again analyzed by radioactive counters.

Working alongside these three strictly biological experiments was a fourth, closely related experiment which analyzed the surface soil with a device known as a gas chromatograph mass spectrometer (GCMS for short). This heated a Martian soil sample to drive off gases, and the gases were sorted according to weight for identification by the mass spectrometer. Its purpose was to detect whatever compounds, particularly organic ones, there might be in the Martian soil.

And, of course, there were the television cameras aboard each lander, which were capable of making the simplest and most positive identification of Martian life—by photographing whatever was growing or moving on the surface.

The Viking missions got under way from Cape Canaveral, Florida, when a Titan-Centaur rocket boosted Viking 1 into space on August 20, 1975. Viking 2 followed on September 9. Both Vikings were identical, and each came in two halves: an orbiter with cameras and other instruments to survey the planet from above, and the all-important lander, sterilized and hermetically sealed in a shell to prevent contamination by terrestrial organisms before launching.

A series of landing sites had been selected by scientists from the Mariner 9 survey of Mars in 1971–72. These sites had to be lowlands, so that the landing parachutes could give the maximum braking effect, and also because the warmest temperatures and the maximum amount of water were likely to be found in lowland areas. But the landing sites also had to be as flat and smooth as possible, because rocky terrain or steep slopes could tip over the tiny lander. Therefore the Vikings took a further close look at the target sites from orbit before releasing their lander craft.

After its ten-month voyage, Viking 1 went into orbit around Mars on June 19 and immediately began surveying its projected landing site in the area known as Chryse (rhymes with "icy"). Chryse is a lowland drainage basin into which water was believed to have flooded during the wet times on Mars. Viking's first photographs from orbit, more detailed than those of Mariner 9, confirmed that water had indeed flowed over the Chryse area, but they also showed that the intended landing site was far rougher than had been anticipated, with many tiny craters and eroded channels not seen on the Mariner 9 photographs. Scientists canceled the landing, originally planned for July 4, the bicentennial of the United States, and began to search for a smoother, safer location.

An alternative site, northwest of the original target area, was tentatively selected, but radar reflections from Earth showed that this was also too rough for safety (reflections of radar beams can reveal the texture of a surface on a scale too small for the orbiter's cameras to see). Eventually a site was chosen farther northwest, about 900 kilometers from the original target area but still in the Chryse basin. Viking 1's lander touched down successfully at 22.27°N, 47.94°W, on July 20, 1976, seven years to the day after the first manned landing on the Moon.

Viking 1's orbiter had also begun to scan the intended landing area for Viking 2, which was by then approaching Mars. It found that the intended site was again dangerously rough. On August 7 Viking 2 joined its predecessor in orbit around Mars, and together the two probes sought out a safe new landing area. Mission scientists decided on a basin known as Utopia, lower even than Chryse but farther north, in an area covered by frosts during the winter. Viking 2's lander touched down here successfully on September 3, at 47.67°N, 225.71°W—on the other side of the planet almost exactly opposite the Viking 1 lander. Two craft were on Mars, making the first on-the-spot search for life on another planet.

One important clue to the possible existence of life on Mars came within hours of touchdown from the instruments that analyzed the composition of the planet's atmosphere. In addition to the 95 percent carbon dioxide already known to exist, the Viking landers detected 2.5 percent nitrogen in the atmosphere of Mars; the remaining few percent was made up of traces of argon, oxygen, carbon monoxide, and one or two other gases. This was the first time that nitrogen had been identified in the Martian atmosphere, and it meant that, along with carbon and water, all the basic ingredients for life had been found to exist on the planet. Added to the fact that the Martian atmosphere seems to have been denser in the past, when the volcanoes erupted and water flowed, prospects for Martian biology looked better than ever.

What was there to see on Mars? Viking lander 1 began to survey its surroundings within minutes of touchdown. After transmitting a picture of one of its footpads resting gently on the stony surface, the lander looked up toward the horizon, about 3 kilometers away. The scene, consisting of scattered rocks and sand dunes, resembled nothing so much as one of the stony deserts in the southwest United States. Color photos from the Viking landers confirmed that Mars is indeed a very red planet. Not only are the surface rocks rust-red, a result of the large amounts of iron oxide they contain, but the sky itself is pink, because of fine dust caught up by winds and suspended in the atmosphere. Yet of life there was no sign.

I, for one, was disappointed not to see a saguaro cactus or two dotting the landscape, along with an occasional Martian equivalent of a Joshua tree. Optimistic biologists had suggested before the mission that organisms large enough to be easily visible to the lander cameras might exist on the surface of Mars. These organisms, the speculation ran, would obtain water that was trapped in the rocks, or they would melt frost; even a thin nightly layer of frost would contain enough water to support a cover of moss or lichen over much of the surface. Large organisms would keep warm more easily in the freezing temperatures than small organisms. One suggested possibility was that the Martian plants might have developed deep roots down to the permafrost layer below the topsoil, and have silica domes like sunshades to ward off the Sun's ultraviolet light. Shells from dead organisms, or their fossils, would have remained visible to the lander cameras even if such life had died out. Alas, the first photographs disproved these daring speculations.

Some indication of the harsh conditions that any Martian creatures would have to withstand was provided by Viking's meteorology instruments, which measured a maximum air temperature in Chryse of −29°C in the mid-afternoon, falling to a low of −85°C at dawn—and

The labeled-release experiment also assumed that Martian organisms need the addition of moisture and some nutrients to aid their growth. A solution of seven simple organic substances, including the amino acids glycine and alanine, was injected into the Martian soil in the test cell. The carbon in the nutrients was radioactively labeled, and the gas above the soil sample was periodically monitored in the hope that any Martian bugs feeding off the nutrient solution would give themselves away by releasing gas containing the radioactively labeled carbon.

As with the gas-exchange experiment, results were immediate and electrifying: on injection of the nutrient into the first Chryse sample, there was a sudden and definite emission of radioactive gas into the chamber, resembling that displayed by biologically poor but nonetheless life-bearing terrestrial soils from the Antarctic.

However, within a day the increase slowed down and then leveled off, remaining almost contant for five sols. Seven sols after the first injection of nutrient, a second dose was added. The overall effect was a drop in radioactive gas in the chamber, which then slowly climbed again. Could this be the first definite sign of life on Mars? One way to find out was to run the experiment again with a control sample that had been heat sterilized to kill any organisms that might be present. So the test cell was cleaned out and a second sample introduced, which was heated to 160°C for three hours. Injection of nutrients into this heat-sterilized sample produced an entirely different response from before, one like that which had been obtained from sterile lunar soils in pre-flight tests on Earth: after a small initial peak, radioactivity fell away to the level of background noise, and even a second injection of nutrient failed to provoke a response. Whatever was causing the reaction had therefore been eliminated by the heat treatment.

This looked very good for Martian biology, although chemical reactions were by no means ruled out. A third trial run remained to be done to clinch the matter: if, over a long period of incubation, radioactive gas continued to be produced at an increasing rate, this would argue for the growth of Martian microorganisms. Gas production from a chemical reaction, on the other hand, would tend to flatten out with time. A third soil sample was therefore tested at the Chryse site in the labeled-release experiment. This sample behaved similarly to the first sample after its first and second injections of nutrients; then a third dose was added, and experimenters sat back to watch the results. Again, the addition of more nutrient led to a drop in radioactive gas, which then began to climb. After a total incubation time of sixty sols, the radioactive gas had returned to its former level, where it seemed to remain constant. There was no sign of the hoped-for increasing rate of gas evolution that would have been a telltale sign of Martian life.

Two samples tested at the second landing site, Utopia, showed responses similar to the first and third Chryse samples. At Utopia, a sample was also subjected to heat treatment before incubation, but this time only to 50°C—not enough to kill all organisms had it been a sample of terrestrial soil, but vastly greater than anything that Martian bugs ever encounter. This so-called cold sterilization led to a reduction by about half in the output of radioactive gas.

So what can we conclude from this experiment? Experimenters Gilbert Levin and Patricia Ann Straat noted that the results of the labeled-release experiment are consistent with a biological interpretation, although they also discussed possible chemical explanations. We must remember the similar sudden response of the gas-exchange experiment to the introduction of water and nutrient. Biologist Cyril Ponnamperuma of the University of Maryland was quick to point out a possible chemical explanation for the results of both the gas-exchange and labeled-release tests, resulting from the highly oxidized state of the Martian soil; both experiments were probably measuring different aspects of the same chemical reactions between the injected nutrient and the Martian surface.

Probably what was happening in the labeled-release experiment was that formic acid (chemical formula HCOOH) in the solution was reacting with the hydrogen peroxide in the soil to produce carbon dioxide, which was the gas being detected by the radioactive counters, and water, which contained no radioactive carbon and thus went undetected. The decrease in gas after subsequent injections of nutrient is explained by absorption of the carbon dioxide in the moist nutrient. It should be remembered that the reaction was not eliminated by heat treatment at 50°C, as would have been expected if Martian bugs were the cause of the gas production. Martian organisms would not previously have encountered such a high temperature and should presumably have been damaged or killed by it, unless they were highly resistant to sudden elevated temperatures as well as to continual sub-zero ones. Therefore, although it must be admitted that the results of the labeled-release experiment are ambiguous, it seems that they can be adequately explained by chemical reactions, without recourse to the supposition of Martian microlife.

Biology test three, the pyrolitic-release experiment, was designed to operate under conditions closely matching those on Mars, and thus might have been expected to give the most reliable clues to the existence of Martian biology. In fact, its results are the most complex and most difficult of all to explain in terms of chemistry. Nevertheless, it seems that these results are not due to Martian biology.

In this experiment, also known as the carbon-assimilation experiment, a sample of Martian atmosphere was trapped above the soil sample in the test chamber. To this genuine atmosphere was added some radioactively labeled carbon dioxide and carbon monoxide; any carbon from the atmosphere taken up by the soil, for instance by living things, would be detected by its radioactivity. A lamp illuminated the test chamber with simulated sunlight; small amounts of moisture could be added to the soil if required, but no nutrients, thereby keeping conditions in the test chamber as Mars-like as possible. The main departure from conditions on the surface was that heat from the spacecraft kept the soil sample around terrestrial room temperature, considerably warmer than the sub-zero ground temperatures normally experienced.

The soil sample was incubated under simulated Martian conditions for 120 hours, then the soil was heated to drive off any carbon-containing compounds, which would be detected by a radioactive detector. The first test at the Chryse site gave a positive response—not as strong as that from terrestrial soils, but positive nonetheless. Something in the Martian soil was fixing radioactive carbon from the atmosphere. A control test was then run, sterilizing the soil at 175°C for three hours before beginning incubation. Any living organisms in the soil should have been completely destroyed by this heat treatment; but the uptake of radioactive carbon did not completely cease. It sank to 10 percent of its previous level, significantly higher than expected from a sterile soil. Whatever was causing the reaction was clearly affected by heat, but not so much as one would have expected for biological organisms.

A third test run at Chryse showed another positive result, although less pronounced than on the first run; the first Chryse sample turned out to have produced the greatest response of all the Viking pyrolitic-release tests at either site. One contributory factor might have been the ground temperature: the first Chryse sample was collected at 7 A.M. local time, as against 11:20 A.M. for the third sample, when the ground temperature was 60°C higher.

In all, six pyrolitic-release tests were conducted at the Chryse site. Chryse sample four produced a response slightly higher than that of sample three. Sample five was heat-treated to 90°C for two hours before incubation, which should have been sufficient to kill off any microorganisms in the soil; but the carbon uptake of the soil was unaffected. Chryse sample six had water vapor added, but produced a level of carbon uptake similar to samples four and five.

At Utopia three pyrolitic-release tests were successfully conducted, two being performed without the light source switched on; one of these runs in the dark gave a positive result, showing that the existence of light is not essential for the reaction to occur. The second Utopia sample to be incubated in dark, dry conditions was taken from underneath a rock, and did not absorb carbon. One Utopia sample was incubated in the light with added water, but for some reason produced virtually zero response.

So what is the explanation here? According to pyrolitic-release experimenter Norman Horowitz and his colleagues, the fact that uptake of carbon by Chryse sample five was undiminished by heat treatment at 90°C, and the failure of sterilizing at 175°C to completely eliminate the reaction in sample two, argues strongly against a biological explanation. Differences in response between different unsterilized samples are probably due to differences in their composition. One significant factor may be their history of exposure at the surface: perhaps only surface material recently exposed to solar radiation can fix carbon from the atmosphere. Even so, as of early 1978 a convincing chemical explanation for the reactions had not emerged, although attempts were being made to simulate the pyrolitic-release results with synthetic Martian soils.

In discussing the possibility of life on Mars, we must take into account the results from a fourth Viking experiment which, while not strictly concerned with biology, is an important adjunct to the three biology experiments. This is the gas chromatograph mass spectrometer

(GCMS), which analyzed the soil for the presence of organic compounds—that is, compounds containing carbon, the basic constituent of life as we know it. The search for organic molecules is important because it provides information on the composition of Martian organisms, their food, and the products of their decay.

A total of four soil samples was studied by this instrument, two at each site. Each sample was heated to drive off gases, which were analyzed according to their weight. On heating to 200°C, carbon dioxide was given off; with further heating to 500°C, oxygen was given off as well. No organic compounds from the soil were detected at either site.

This failure to detect organic material on Mars has caused considerable surprise. Tests with an identical GCMS on Earth showed that it is sensitive enough to detect organic material in soils from the Antarctic, and even the traces of organic compounds present in meteorites. Since meteorites regularly land on Mars, many scientists had expected that traces of organic molecules from these would have shown up in the GCMS results. The answer seems to be that the organic matter from meteorites is so thoroughly mixed up in the Martian soil that it becomes too diluted to be detected by the GCMS.

Organic compounds would also be expected to be synthesized on Mars from the carbon of its atmosphere by the action of ultraviolet light. But ultraviolet light can break up molecules as well as produce them, particularly in the presence of oxides like those on the Martian surface. The results from the GCMS show that if organic compounds are produced at all on the surface of Mars, they must be broken up just as quickly as they are formed.

Although the GCMS is sensitive, a small population of Martian microorganisms could still exist in the soil and go undetected; for instance, the amount of carbon fixed by the pyrolitic-release experiment, enough to support a few hundred bacteria, would be below the level of sensitivity of the GCMS. But pre-flight tests with samples of biologically poor Antarctic soil have shown that there is 10,000 times as much organic matter in the form of debris in the soil than there is in the bodies of organisms that live in the soil. This debris provides food for the organisms and consists of the remains of dead organisms. But the negative results from the GCMS show that no such level of organic debris exists in the soil of Mars. Therefore, GCMS experimenter Klaus Biemann and his team conclude that organisms based on terrestrial biochemistry are unlikely to exist on Mars unless they are *very* much more efficient at scavenging organic debris than organisms on Earth.

The GCMS results help underline the message of the Viking biology experiments: Mars is not teeming with life. In all, a total of twenty-six experimental tests was carried out on Mars by the two Viking landers, twice as many as anticipated, in an attempt to clarify the problem of whether or not there is life on Mars. The red planet turns out to be frustrating for biologists—it is so nearly favorable for life, but not quite. I am reminded of an old cartoon depicting the return to Earth of the first astronaut to visit Mars. He is besieged by reporters clamoring to know "Is there any life on Mars?" He replies: "There's a little on Saturday night, but it's pretty dull the rest of the week." Likewise, Viking is telling us that, from the point of view of life, Mars is pretty dull right now.

Although it is fair to say that Viking did not discover life on Mars, the possibility of Martian biology cannot yet be completely ruled out. For one thing, the Viking biology experiments were limited in scope; a small population of life forms might exist that did not respond to the Viking biology experiments. As Viking chief biologist Harold Klein has put it, looking for life on Mars is like fishing in an unknown lake with several different types of bait; perhaps Viking used the wrong bait. Alternatively, a much larger population of organisms might exist at more favored locations, such as around the polar caps. Whatever the case, it is certainly true that Mars does not contain an abundant terrestrial-type biota at the Viking landing sites in Chryse and Utopia.

What of the future? A number of plans have been suggested for future exploration of Mars, although nothing has been decided as of this writing. The simplest plan, a straight repeat of Viking with new biology experiments, does not find much favor. A better idea might be a Mars rover, similar to Viking but on caterpillar tracks so that it can be steered around in search of the most interesting sites to examine. Another suggestion is a Mars surface penetrator, a long and pointed instrumented package that parachutes to the surface at high speed, burying itself several meters into the ground to analyze the subsurface composition and to sense Marsquakes. Such a surface penetrator would probably be dropped from a Mars orbiter.

Most exciting of all, though, is a Mars sample return mission—actually bringing back soil from Mars by an automatic vehicle like the Soviet lunar landers which have automatically returned samples of moonsoil. Perhaps only in this way can we finally answer the question about life on Mars. A modified Viking could be sent to the surface of Mars, scoop up a sample of surface material, and feed it into a return capsule which blasts off for the journey back to Earth. Such a mission was strongly favored in a National Academy of Sciences report on post-Viking biological investigation of Mars published in 1977. Although technically possible, a Mars sample return is unlikely to take place until 1990 at the earliest.

And even if there is no life on Mars today, we will eventually put life on the planet. Men will one day stand on Mars, although that is going to be such a difficult and expensive undertaking that no one can tell when it is likely to happen—certainly not before the turn of the century. A summer study at NASA's Ames Research Center in 1975 considered the possibility of making Mars fit for human habitation—that is, engineering the planet so that it had free oxygen, water, and a bearable climate.

Such an adventurous piece of planetary engineering could be accomplished by seeding the planet with ultraviolet-resistant bugs which would break down the carbon dioxide atmosphere to produce free oxygen. But it would take 100,000 years to make Mars habitable in this way. An alternative is to warm up Mars by increasing its absorption of sunlight. This can be done by decreasing the reflectance of its polar caps, either by planting vegetation on them or by sprinkling dark dust over them. The extra heat absorbed as a result of this treatment would vaporize the polar caps, increasing the density of the atmosphere, melting the permafrost under the surface, and producing a habitable environment. Even so, the new climate might still not become stable for 10,000 to 100,000 years. We might reduce this time scale by bioengineering a supermicrobe capable of a vastly improved rate of photosynthesis, or we could artificially pump more heat into the polar caps to increase their rate of evaporation.

But that is far into the future. In the meantime, let us turn toward some more readily attainable goals.

REFERENCES

Biemann, K., et al., *Science*, Vol. 194 (1976), p. 72.
——— *Journal of Geophysical Research*, Vol. 82 (1977), p. 4641.
Horowitz, N. H., et al., *Science*, Vol. 194 (1976), p. 1321.
——— *Journal of Geophysical Research*, Vol. 82 (1977), p. 4659.
——— *Scientific American*, November 1977, p. 52.
Klein, H. P., et al., *Science*, Vol. 194 (1976), p. 99.
Klein, H. P., *Journal of Geophysical Research*, Vol. 82 (1977), p. 4677.
Levin, G. V., and P. A. Straat, *Science*, Vol. 194 (1976), p. 1322.
——— *Journal of Geophysical Research*, Vol. 82 (1977), p. 4663.
Oyama, V. I., et al., *Nature*, Vol. 265 (1977), p. 110.
Oyama, V. I., and B. J. Berdahl, *Journal of Geophysical Research*, Vol. 82 (1977), p. 4669.
Ponnamperuma, C., et al., *Science*, Vol. 197 (1977), p. 455.

5

Living Out There

Colonies in space have been a dream of mankind since the time of the Soviet space prophet Konstantin Tsiolkovsky, who in an 1895 science-fiction story, "Dreams of Earth and Heaven," visualized what we would now term a space station orbiting Earth. Tsiolkovsky went further in his 1903 book *The Rocket into Cosmic Space,* introducing the idea of spinning the station to provide artificial gravity, and using sunlight for power. He and his readers felt the lure of living in space, and recognized the promise of new freedom that it afforded. Now, we are ready to take those first steps that may lead to mankind's eventual colonization of the Galaxy.

Since Tsiolkovky's work was little known outside Russia, impetus for space station studies in Europe came from the German spaceflight pioneer Hermann Oberth, who in his small but influential book *The Rocket into Interplanetary Space,* published in 1923, envisaged the use of manned stations for observing the Earth and heavens, as well as bases for construction and refueling of spacecraft in orbit. Hermann Noordung (the pen name of an Austrian army captain named Potocnik) in 1929 proposed a wheel-shaped station spinning to provide gravity, an idea which sounds very familiar today. Noordung's wheel was 100 feet in diameter and carried large mirrors to focus sunlight to provide power. In 1951, Wernher von Braun designed a larger wheel-shaped station built in orbit from prefabricated sections. This concept was widely publicized, and from then on the whole idea was up for grabs.

For instance, Dandridge Cole in 1963 envisaged hollowing out an asteroid to create living space. Arthur C. Clarke in *A Fall of Moondust* (1961) pointed out the advantages of the gravitationally stable Lagrangian points in the Moon's orbit as locations for space stations. The idea of building giant cylinders in space, 1 kilometer long and 300 meters in diameter, to provide living space for up to 20,000 people was broached in a daring 1956 proposal by Darrell Romick.

These speculations lay around like prefabricated components awaiting assembly into a coherent whole when in 1969 Princeton physicist Gerard K. O'Neill realized that not only was space colonization technically feasible within our lifetime, but that it had strong merits which made it highly worthwhile. Since then, colonies in space have beckoned as the next major goal once the Space Shuttle transportation system is in operation. Beyond that, space colonization (or, as O'Neill terms it, the "humanization" of space) has major implications for interstellar travel and communications—both by us and by any other civilizations that may exist. In looking at space colonies we may be seeing the genesis of manned starships and space communication bases.

O'Neill's interest in space colonization began by chance at a seminar for a first-year Princeton physics class. O'Neill is a high-energy physicist, noted for his invention of the storage-ring technique for increasing the collision energies of atomic beams from particle accelerators. His imagination had already been caught by the opportunities of space; in 1966 he applied as a scientist-astronaut for the continuing program of space science that NASA was then planning. But that program was cut in 1967 just before the final class of new astronauts was selected, and O'Neill remained at Princeton. There are no regrets, as he explained to me in an interview for the *New Scientist*: "I don't think that any of the things I could have done as a scientist-astronaut would have made nearly as much of a contribution as I can hope to make with the humanization of space."

His attempt to make physics relevant on a wider scale in the face of the apparent disillusionment with science in the late 1960s led to that 1969 physics seminar at which he set his class the question: Is a planetary surface the right place for an expanding technological civilization? The answer, it soon emerged, was no. The surface areas of the Moon and Mars, for instance, would no more than double the available land area of Earth, and of course their environments would need to be radically engineered to be suitable for life. These limitations do not apply to an enclosed colony in free space.

There are three basic shapes for a colony rotating to provide gravity: a wheel, a sphere, or a cylinder. O'Neill's group felt that a wheel was more appropriate for a space station than for the mini-worlds of which they were thinking. This left the cylinder as the best choice, for it has the maximum inner surface at full artificial gravity (in the sphere, gravity falls off as the sphere curves in toward its rotation axis). Rough calculations of mechanical strengths soon showed that cylindrical colonies could be built with diameters of several kilometers, offering enough land area to support millions of people. The colonies would be constructed of materials mined from the Moon. They would be heated and illuminated by sunlight reflected in through large windows by mirrors on the outside. Eventually, using asteroids for new building material, it would be possible to construct enough colonies to provide for 20,000 times Earth's current population.

Fascinated by now, O'Neill continued the study in his spare time. He quickly saw the possibilities of providing pleasant living conditions without the problems of pollution, overcrowding, food shortages, and energy crisis that mar our technological age on Earth. Immediately, the stifling Limits to Growth philosophy, which predicted gloom and doom for our technological future, was seen to be undermined by the opportunities of space—the very enterprise which many Limits to Growth supporters criticized as the most wasteful and irrelevant of all.

These utopian possibilities quickly attracted a following on the college circuit as O'Neill lectured about his ideas. With some supporters, he held a modest conference at Princeton on space colonization in May 1974. In September 1974 the magazine *Physics Today* published an article by O'Neill outlining his proposals, which other magazines had dismissed as little short of science fiction. The response from scientists and public alike was immediate and positive.

It became apparent from that first Princeton conference that there were no insuperable technological obstacles to building the colonies, if we so wished, by the turn of the century; all that was needed was a commitment. But why should we build space communities? What would be the economic return that would make investment in them worthwhile?

At the end of 1974, with the feasibility of the space-colony concept secure, O'Neill began to consider the possibility of using his colonies as construction bases for the satellite solar power stations proposed several years earlier by Dr. Peter Glaser of Arthur D. Little, Inc., of Cambridge, Massachusetts. Each solar power satellite would produce 10,000 megawatts, enough to power a city. In geostationary orbit around the Earth, these satellites would collect sunlight via solar panels, turn it into microwaves (short-wavelength radio waves), and beam it back to receivers on Earth where it would be converted into electricity. In an experiment to test this technique, a bank of seventeen lamps was lit by a microwave beam of 30 kilowatts power transmitted across one mile from the 26-meter NASA tracking dish at Goldstone, California.

An alternative to collecting sunlight with solar cells is the approach of Gordon Woodcock and Daniel Gregory of the Boeing Aerospace Company, who have considered using sunlight as a heat source to drive gas turbines, which generate the power that is beamed back to Earth in the form of microwaves. A turbine-powered satellite may be more readily constructed than the version using solar cells envisaged by Peter Glaser, because the silicon crystals of solar cells are both expensive and intricate to make—that's why they have not already displaced traditional power sources on Earth. Cheaper and simpler solar cells are on the way, but they may never be competitive with turbine-generated electricity, either on Earth or in space.

The advantage of placing solar power collectors in orbit is that sunlight in space is at least four times as plentiful as at the sunniest places on Earth. In space, the Sun never sets, and there are to interruptions from bad weather. Sunlight seems the ideal energy source, because it is clean, endless, and free—apart, that is, from the cost of harnessing it. Solar power satellites have the drawback that they are large and heavy, which would make them expensive to launch from Earth. But with an O'Neill colony there are no launching costs, because the space colonists build the power stations in space using the same techniques as used to build the colonies themselves. They then sell the electricity to Earth.

The whole economics of satellite solar power stations is therefore transformed, as O'Neill reported in the magazine *Science* in December 1975. He noted the enormous potential market for electricity in the United States, where it is estimated that around 100,000 megawatts a year of new generating capacity will be needed by the year 2000. Building power stations on Earth to meet such demand will require investment of several hundred billion dollars, and the environmental threat from such a profusion of new power plants, coal-fired or nuclear, is itself alarming. Solar power from orbit not only has environmental advantages, but in the long run it will also be cheaper than building power stations on Earth.

A rough estimate of the total outlay to produce the first space colony which will start building power satellites is $100 billion. That first colony will be under construction for six years or so, and the first 10,000-megawatt power satellite will come into operation one to two years later. More colonies will then be built, which will produce their own power satellites, so that the enterprise becomes self-perpetuating without further help from Earth. Thirteen years after the start of the project, the power satellites could be supplying sufficient energy to meet the U.S. annual need for new generating capacity.

To ensure market penetration, the electricity must be competitively priced. O'Neill assumes a starting price of 1.5 cents per kilowatt-hour, equivalent to the lowest-cost electricity currently available in the United States, which comes from nuclear reactors. This price would be progressively dropped until solar power satellite electricity was by far the cheapest available. Even at these low rates, after twenty-four years the whole project will have paid for itself, and be returning handsome profits.

Solar power from space thus appears to be an attainable and desirable solution to our energy crisis. If the economic and technical arguments stand up to scrutiny, the colonies will certainly be built. Once the colonies are in existence, of course, their other advantages—as sites for materials processing and space industrialization, as a jumping-off place for the rest of the solar system, as an astronomy platform, and simply as a desirable place to live—would come as an added bonus. NASA has been sufficiently impressed to fund O'Neill's research since 1975.

O'Neill originally envisaged his first colonies as small cylinders, 1 kilometer long and 200 meters wide, spinning three times a minute to provide Earth-normal gravity. For stability, the cylinders would be linked in pairs, rotating in opposite directions: they must be kept with their long axes oriented toward the Sun. As many as 10,000 people could be housed in the first colony. Once this was set up, its occupants would provide the workforce for building new and larger models, pairs of cylinders up to 32 kilometers long and 6.4 kilometers in diameter, with space for millions of people.

O'Neill foresaw that the colonies would be established at the gravitationally stable Lagrangian points of the Moon's orbit. As seen from Earth, these lie 60° ahead of and behind the Moon, and are known as L4 and L5 respectively. These are the points at which the gravitational effects of the Earth and Moon exactly cancel, as calculated in 1772 by the French mathematician Joseph Louis Lagrange. Highly complex calculations performed by computer in 1968 showed that the additional gravitational effects of the Sun destroyed the stability of these theoretical points, but replaced them with something at least as good: stable kidney-shaped orbits centered on L4 and L5, along which a colony would take eighty-nine days to travel. There is room on each orbit for thousands of colonies, so the situation is actually better than if we were stuck with L4 and L5 alone.*

Apart from their stability, the areas at L4 and L5 have the advantage that they are seldom eclipsed by the Earth, and they are easily reached from the Moon's surface, which is where the construction material will be mined. The vital knowledge without which the colonization concept could not have got under way was that of the chemical composition of the Moon, as revealed by the samples brought back from the Apollo missions. It turned out that the Moon is surprisingly rich in aluminum, titanium, and iron; it is from these metals that the colonies will be built. There is also plentiful silicon, which will be used for the glass of the colonies' windows, and abundant oxygen, which will serve to supply an atmosphere and also as a rocket fuel. Lacking in Moon rock are carbon, nitrogen, and hydrogen, which will initially need to be brought from Earth but will eventually be obtained from asteroids.

* There are three other Lagrangian points, one behind the Moon as seen from Earth, one between the Moon and Earth, and one on the opposite side of Earth from the Moon.

A mining team of perhaps 150 men will be located on the Moon with bulldozers and scoops, probably on the southeastern Mare Tranquillitatis, where they will scrape up and shoot into space a million tons of lunar topsoil a year. Even a generation of digging will not produce a crater large enough to be seen from Earth. It is not a rocket that will propel the lunar material into space, but a device called a mass driver, which uses the technology developed for magnetically levitated trains. Parcels of lunar soil are fitted into so-called buckets, which are then accelerated along a guideway by magnetic fields at 100 times the force of gravity (100 g) to the escape velocity of the Moon, 2.4 km/sec. The bucket shoots out from the end of the track and is in free flight, where its trajectory is checked by laser beams and is corrected if necessary by magnetic impulses. Then the bucket enters another section of track where it is rapidly decelerated, allowing the payload to shoot out into space. As the payload curves away from the Moon, the bucket is ferried back to the starting point many kilometers away for reloading. A model mass driver with an acceleration of 35 g was demonstrated at the third Princeton conference on space manufacturing in May 1977; models with greater performance have since been developed.

Two payloads will be launched each second. After two days' flight through space they will reach L2, the Lagrangian point 64,000 kilometers beyond the Moon as seen from Earth. Here the parcels of Moonstuff will be caught by a rotating bag made of Kevlar, the synthetic material used for bulletproof vests. When full, the catcher moves off to the colony site, where the ore-processing plant will reduce the precious lunar material into metal and glass for construction.

In addition to O'Neill's yearly Princeton conferences, NASA has held a number of summer studies at the Ames Research Center to examine the space colonization concept. At the second of these, in 1976, it was realized that the L4 and L5 Lagrangian points might not after all be the best locations for the first colonies. A high Earth orbit, two-thirds of the way from Earth to the Moon, now seems as good a location as any; it is easier than L4 and L5 to reach from both L2 and Earth, and it would also be easier to deliver power satellites from there to geostationary orbit. This discovery was somewhat embarrassing, for the L5 location in particular had become closely associated with O'Neill's vision; a grass-roots organization formed to support O'Neill's space colonization concept has styled itself the L5 Society. But the large, primarily residential colonies foreseen by O'Neill may yet be stationed at L4 and L5.

The first NASA summer study, in 1975, had also led to a revision of ideas about the shape of that first colony. Whereas O'Neill had been thinking in terms of small cylinders that would need to rotate three times a minute to provide Earth-normal gravity, a medical student named Larry Winkler maintained that such a speed of rotation would produce motion sickness in the occupants; a rotation rate of one revolution per minute was the maximum he considered allowable. A cylinder shape was still acceptable for the largest colonies, capable of housing millions of people, because they would need to rotate at only 1 rpm or less to simulate Earth gravity. But in the case of the small first colonies, the occupants would either have to be specially selected to withstand motion sickness, or they would have to put up with a slower spin that gave them less than Earth gravity. Or the colony would need to be redesigned.

The greater the diameter of a colony, the slower it need spin to simulate Earth gravity, so the 1975 summer-study participants looked again at wheel-shaped designs and came up with a concept that was dubbed the Stanford torus. This was a ring like a bicycle tire, 1.8 kilometers in diameter; the tire tube itself, inside which the 10,000 colonists live, is 130 meters across. The torus rotates once a minute to provide Earth-normal gravity on the inside surface of the tire rim. Docking ports for arriving spacecraft are situated at the station's hub, whence passengers transfer to the outer ring along spokelike corridors. At the hub, gravity is virtually nonexistent because the rotational forces are so low. Here would be the station's main recreational area, in which the inhabitants could experience the delights of low-gravity living, including human-powered flight. Attached to the hub by transfer tubes are the manufacturing areas of the colony, where space freighters will unload their cargoes of lunar materials to be processed into the metal tubes, girders, and panels for building the power satellites that govern the colony's economy.

Unlike an O'Neill cylinder, whose rotation axis is aligned toward the Sun, the axis of the Stanford torus points at right angles to the Sun. Above the colony, opposite the manufacturing areas, hovers a giant mirror a kilometer or so across, angled at 45° to reflect sunlight down onto the colony. A second set of mirrors around the hub reflects the light into the torus; tilting these mirrors up and down produces the effect of day and night. To the inhabitants of the colony, who would have their feet on the outer rim of the torus, sunlight would appear to come from overhead.

O'Neill, though, preferred to forge a compromise between the lower recommended rotation rate and his favored colony shape. He rejected the Stanford torus design; for the 10,000-inhabitant Island One, as it became called, he suggested instead a sphere 460 meters in diameter, rotating twice a minute to give Earth-normal gravity at its equator. This is the design that has been widely illustrated and is sometimes referred to as a Bernal sphere, after the Irish scientist J. D. Bernal, who foresaw the construction of "space arks" for space colonization. Gravity on the inner surface of a spinning sphere falls off toward the axis of rotation, so most of the human habitation will be in the equatorial region. Around the equator, a river may meander.

If Island One is made of an aluminum shell, the thickness of the sphere will vary from a maximum of about 18 centimeters at the equator to 5 centimeters near the rotational axis. Since the gravitational stresses are least near the axis, that will be the best place for the windows. There will be a ring of windows around the axis at each end of the sphere, but they will not look out directly into space; instead, they will be surrounded by a circular mirror which deflects sunlight inside. Agriculture will be carried on in areas at the ends of the sphere, surrounding the docking ports on the axis of rotation, while the industrial regions will be entirely separate, reached by commuter ships. All power for the colony will be supplied by an on-board solar power station.

Whatever the shape of Island One, or wherever it is eventually located, one thing is clear: construction will not begin until considerable experience has been gained in living and working in space via the Space Shuttle, the reusable space transportation system due to come into full-time operation in 1980. According to current NASA plans, the Shuttle will probably be used in the mid-1980s to establish a full-scale space station, far more impressive than the makeshift Skylab of 1973–74. Although it is too early to be sure exactly what such an interim station would look like, it will probably be made from individual units that plug together, and could house up to 200 people. The experience in space industrialization gained through such a station would be an essential stepping stone to O'Neill's space colonies, which could begin to dot the sky like faint new stars between the years 1990 and 2000.

Even 200 men, though, will not be enough to build the first space colonies. Before Island One can begin to take shape, we would need to set up an industrial facility, colloquially termed the construction shack, containing a workforce of perhaps 2,000, who would process the lunar ores fired into space. For the workers in the construction shack and the lunar colony, life will be rather like that on an offshore oil rig: hard and monotonous, and perhaps hazardous. Room will be at a premium, for these initial structures will need to be delivered into space from Earth. One proposed design foresees the shack as a sphere 100 meters in diameter. It would need to contain rolling mills and casting beds to produce the metal sheets and cables required for the construction of Island One and the first satellite power stations. Its total weight might be 10,000 tons; a similar weight of material might need to be delivered to the lunar surface to set up the first mining base. In addition to these materials, there will also be many thousands of colonists requiring transport into space.

Even the modern Space Shuttle is too small and too expensive to cope with the job without adding considerably to the space colonization budget. What is needed is a still larger and more economical launcher, termed the Heavy Lift Launch Vehicle (HLLV). Various studies based on an extension of the current Shuttle system or even a reusable version of the Saturn V first stage show that although such launchers may take several billion dollars to develop, they will still be worth it: launch costs make up the major part (perhaps 60 percent) of the $100 billion estimated for the total colonization project, which is about three times the cost of the Apollo program at current prices. Without the new vehicles, the colonization cost will be almost doubled. Nevertheless, the need to develop them so soon after the introduction of the Shuttle may be a political stumbling block to space colonization. So would the frequency of launches: one a day for several years, which is considerable for a spaceport, though trivial in comparison with aircraft movements.

One of the dangers to the workers in the construction shack will be radiation damage. The walls of the shack can be shielded to exclude most cosmic rays, but every so often flares erupt on the surface of the Sun, spewing out beams of radiation strong enough to kill a man. The construction-shack workers will then have to retreat to lead-lined flare shelters for a day or so until the danger is past. But what about the colonies themselves? Many people assume that long-term exposure to radiation and the chances of meteorite impact will make living in space unacceptably hazardous.

It turns out that radiation can easily be screened out, by surrounding the colonies with soil two meters thick. For the Stanford torus, 10 million tons of shielding will be required; the Bernal sphere has a much smaller surface area and would require only one-third as much shielding mass, which is another reason for favoring it. These figures compare with a mass of half a million tons for the colony structure itself. In either case, the shielding will be nothing more glamorous than the slag left over from industrial processing of the lunar soil, and so will not require to be sent up specially from the Moon. Only the window openings present a problem; they must be shielded from a direct view of space by the angled mirrors that reflect light into them. In the case of the largest colonies, the giant cylinders of O'Neill's grandest vision, the thickness of the cylinder structure and the depth of the atmosphere will be sufficient in themselves to absorb cosmic-ray particles without further shielding. Thus, space colonists in all sizes of habitats can live as safe from radiation as people on Earth.

Meteoroid damage will be more of a problem, but it will not be the catastrophe one might instinctively guess. Most meteoroids are so small that they would have little more than a gentle sandblasting effect on the outer skin of the colony, while many of the larger ones are so fragile that they would break up on impact like clods of soil. (Most of the meteorites that encounter the Earth do not survive the passage through the atmosphere.) The thick cosmic-ray shielding of the early colonies will also serve to absorb meteorite impacts. In the largest colonies, which will have plentiful window areas, glass panels will probably be broken every few years by meteorites of around 100 grams weight. Since the panes of glass will be similar in size to those in domestic windows, leakage will be slight: loss of one window panel in a large colony would lead to complete decompression only after many years, which gives repair teams plenty of time to patch the leak. Meteorites weighing a ton or more, capable of punching a large hole in the shell, will be encountered every few million years, so space colonization will be less hazardous than most other forms of human activity.

Everything about a space colony will be exotic, including its interior. Colonies will control their temperature by regulating the amount of sunlight they allow inside; O'Neill imagines that Island One might be given a tropical climate like that of Hawaii. Tests on moonsoil have shown that plants grow very well in it, with the addition of water and nitrates, so there would be plentiful vegetation in the colonies—trees, bushes, and attractive flowers. Gardeners will not be troubled by weeds or pests, because these will be left behind. Strict regulations will prevent the importation of undesirable parasites and diseases.

Despite the projected population for Island One, the inside area of the sphere is so great that there will be ample room for everyone. Building materials will be bricks cast from slag, metal extracted from lunar soil, and glass made from the silicon which is abundant in the lunar crust. Therefore apartments and houses in the colonies will consist of much the same materials as in modern buildings on Earth.

Occupants looking out from their apartment or garden in Island One will see the sphere curving away on all sides. Trees, buildings, and people will jut out at the horizontal up the sides of the sphere and appear to hang upside down on the opposite side of the sphere, 460 meters overhead. As one walked uphill to the rotation axis, one would feel lighter as the effect of artificial gravity diminished. At the axis itself one would be weightless; here would be the low-gravity swimming pool, and from here the high-fliers of the colony would launch themselves into the air to attain man's dream of human-powered flight.

Atmosphere, water, and food are the prime requirements of any colony. Many people assume that the colonies must be regularly supplied with consumables from Earth, but this is not so. Each colony will be a mini-Earth, with its own closed ecological system. Wastes will be recycled, while plants will remove carbon dioxide and replenish oxygen; other air-freshening will use the same techniques as in submarines. Water will at first be made by combining hydrogen brought from Earth with oxygen released from the Moon's crust, but eventually other sources of hydrogen will be found—notably the asteroids. Since most of Earth's atmosphere is nitrogen, which is not important for our breathing, atmospheric pressure in the colonies can be reduced without danger. Apollo astronauts, for instance, breathed pure oxygen at a pressure one-third that at sea level. In the colonies, some nitrogen will be mixed with the oxygen to bring up atmospheric pressure to half that at sea level on Earth.

Agriculture in the colonies will actually be easier than on Earth. Farm areas will be kept separate from the living areas, so that their conditions can be tailored to the crops being grown. Since the climate can be controlled, the growing season would be continuous. One estimate is that a colony of 10,000 people could be supplied amply by 100 acres of agricultural area. There will be cereals, fruit, and vegetables. Goats will provide milk, and for meat the most efficient source would be rabbits and chickens. Fish should also be easy to raise in space. Steak will literally be rare at first in the colonies, because cattle are both heavy and inefficient at using feedstuffs. Eventually, though, popular pressure will lead to the introduction of a wider selection of farm animals.

Good though the living will be in Island One, and its sister worlds in space, it will still fall short of the ultimate possibilities of the largest colonies. Once the return from the power satellites has put space colonization on an economic footing early in the next century, attention will be turned toward building the second-generation colonies. Various nations may wish to design and place orders for their own versions. Plausible designs would be a larger Bernal sphere, 1,800 meters in diameter, or a cylinder 3.2 kilometers long and 640 meters in diameter, both capable of housing over 100,000 people. By the middle of the next century the time may be right for Island Three, the very largest colonies of all, housing millions of people in cylinders 32 km long and 6.4 km in diameter and spinning every two minutes. (The largest cylinders that could be built within the physical strength of materials are 120 km long and 24 km in diameter, but it is unlikely that the engineers would want to go so far.) With colonies the size of Island Three, it would be possible to provide new living area at a rate faster than the growth of population on Earth.

Down the length of each cylinder would run three strips of land, alternating with three strips of windows. Opposite each land strip would be a window, through which sunlight would be reflected by a longitudinal mirror outside the colony. Tilting and turning these mirrors each day will give reassuringly familiar impressions of sunrise and sunset. Even more Earth-like, in cylinders of such size the sky will appear blue and clouds will form at around 1,000-meter altitude. There will be rainstorms and a small but definite variety of weather from day to day.

Whereas the first colonies would serve primarily as bases for power-satellite construction workers and for scientists, the larger colonies are seen as permanent habitats for humans. Whole families will move to the colonies in search of new opportunities, like the flood of immigrants to America in the last century. Not only will the colonies offer high living standards in an attractive, pollution-free environment; they will also offer freedom.

Residential colonies will demand a much greater variety of foodstuffs than would be available in Island One. O'Neill suggests that there should be a ring of smaller agricultural cylinders which can be adapted to provide abundant food of all kinds. Unless diseases unexpectedly invade agricultural cylinders (in which case the cylinder will be closed and sterilized) there will be no chance of crop failure, so food will be more abundant than on Earth.

Of course, we must expect problems. Will the moisture in the atmosphere condense on the colony windows? What unexpected ecological effects will arise in the artificial biosphere? Will there be metal corrosion and fatigue as in bridges and aircraft? What will be the political relationship between the colonies and Earth as they become increasingly independent? What, above all, will be people's response to living in an enclosed environment? Space living may suit only a small proportion of humanity.

As the number of large colonies grows, the smaller ones may be refurbished for special uses. For instance, they might be occupied by scientists carrying out dangerous experiments, such as in genetics, that require total isolation; or social groups may take over colonies for experiments in community living. Zero-gravity laboratories at the axes of the colonies could make astounding developments in physics, medicine, or metallurgy, generating further valued income for the colonies. Space colonies are likely to become the scientific mecca of the next century. Astronomers will be fighting to set up large telescopes in the colonies to observe the Universe free of the blurring effects of Earth's atmosphere. One colony, equipped with giant radio telescopes, may become the headquarters of the search for messages from the stars. Another might be adapted as an interstellar spacecraft.

We need not despoil the Moon to build these additional colonies. We can instead use raw materials extracted from asteroids, many of which are believed to contain the hydrogen, nitrogen, and carbon that is missing on the Moon. Although it would require considerable energy to bring back an asteroid from the main belt between Mars and Jupiter, a select group of asteroids have orbits that bring them close to Earth. There may be 100,000 asteroids larger than 100 meters in diameter that pass suitably close to Earth. Astronomer Brian O'Leary of Princeton has calculated that an Earth-approaching asteroid 200 meters across weighing 10 million tons, containing enough material to build ten satellite solar power stations, could be retrieved at a cost no more than half, and perhaps as little as one-tenth, that of bringing the same mass up from the Moon. Many asteroids—the carbonaceous chondrites—are rich in complex hydrocarbons, which the space industrialists can use to make plastics. Eventually the major manufacturing industries of Earth will be forced to move out into space, where raw materials are so much more plentiful.

It will not be many decades before colonies are set up elsewhere in the solar system. With a suitably modified mirror system, O'Neill colonies could function on incident sunlight out to the orbit of Pluto. Beyond about three light-days from the Sun, the light becomes too weak to use without impracticably large mirrors. But with a nuclear power system on board, the colonies could travel through space as self-sufficient interstellar arks.

This is the real, long-term significance of the space colonization program. Not many years after the first colonies have been set up they will be totally independent of Earth. Even if some disaster should eradicate life on Earth, they would be able to carry on. The colonies thus act as a lifeboat for humanity, effectively making mankind as a species immortal. And once mankind has made this first breakout into space, interstellar travel—even colonization of the Galaxy—follows naturally.

The question is: Since that first breakout into space can occur so readily, with even our relatively primitive level of technology, has it happened elsewhere? And if so, where are they?

REFERENCES

Heppenheimer, T. A., *Colonies in Space* (Harrisburg, Pa.: Stackpole Books, 1977).
——— and M. Hopkins, *Astronautics and Aeronautics*, March 1976, p. 58.
O'Leary, B., *Science*, Vol. 197 (1977), p. 363.
O'Neill, G. K., *Science*, Vol. 190 (1975), p. 943.
——— *Astronautics and Aeronautics*, October 1976, p. 20.
——— *The High Frontier* (New York: Morrow, 1977).
——— *Astronautics and Aeronautics*, March 1978.
Ridpath, Ian, *New Scientist*, Vol. 74 (1977), p. 718.
——— *New Scientist*, Vol. 78 (1978), p. 520.
Salkeld, R., *Astronautics and Aeronautics*, September 1975, p. 30.

PART TWO

MESSAGES FROM THE STARS

6

Our First Ship to the Stars

Less than twenty years after the orbiting of the first Sputnik, four probes to the outer planets had been launched—Pioneers 10 and 11, and Voyagers 1 and 2—that will eventually leave the solar system and drift dead and dark among the stars millions of years hence. They each carry messages for any civilization that may intercept them (see Chapter 8). Therefore we already have the ability to reach the stars, albeit slowly. When the Pioneer and Voyager probes leave the solar system they will be traveling at 17 km/sec, faster than any objects before them. But even at that speed they would still take the best part of 80,000 years to reach Alpha Centauri.

The O'Neill colonies described in the previous chapter could act as habitats for long-duration manned interstellar missions, like the space arks of science fiction. But they, too, would provide a very slow boat to the stars. Can we hope to reach the stars within a human lifetime? And, if so, when might it be possible?

These questions have been addressed by a team of thirteen engineers, physicists, and astronomers at the British Interplanetary Society in a design study called Project Daedalus. Their aim was to design an unmanned craft, based on the technology likely to be available by the turn of the century, that could fly to Barnard's star within about fifty years. Barnard's star was chosen as a target because of its high probability of having planets, although there is nothing to stop the probe from being sent instead to other nearby stars. In fact, there could be a whole fleet of Daedalus craft built to examine the Sun's stellar neighborhood. The four-year Project Daedalus design study, which terminated after more than 10,000 man-hours of work in April 1977, proved that it would theoretically be possible to send unmanned probes to the nearest stars, and get back information, within the next century.

Because stars are so far apart, it is a basic fact of the Universe that interstellar missions must always last longer than missions to the planets of our own solar system. Even if our probes could travel at the speed of light, which theory says is impossible, we on Earth would still need to wait twelve years to receive results from Barnard's star—six years for the journey, and another six years for the radio message to come back. In practice, interstellar journeys would be undertaken at considerably less than the speed of light, and so would take several tens of years to complete. Judged in this light, Daedalus is a very impressive prototype starship.

Daedalus makes no appeal to the mythical space warps or other inventions of science fiction to propel it across the yawning gulf to the stars. Rather, it uses the most powerful energy source currently available to man—hydrogen fusion, as in a nuclear bomb. If we can control fusion power, we have the practical answer to interstellar travel. At the moment, our only way of using such power for space travel is in a nuclear pulse rocket, which is the technique adopted for Daedalus, although future developments in fusion reactors may give us improved engines for starships.

Daedalus is a two-stage craft of enormous proportions, 230 meters long and 200 meters across at its widest, weighing 53,200 tons, all but a few thousand tons of which is propellants—bombs in the form of small spheres a few centimeters in diameter. In comparison, the Saturn V Moon rocket stood 111 meters tall and weighed 3,000 tons fully fueled. Daedalus carries a very substantial payload, 450 tons, which is six times the weight of Skylab, the largest and heaviest spacecraft so far launched. Over half the Daedalus payload is taken up by twenty-three sub-probes designed to examine the Barnard's star system and interstellar space en route.

Daedalus will not be built and launched from Earth, but will be constructed in space at the same space industrial facilities as used for the O'Neill colonies described in the previous chapter. Daedalus's structural materials will be mostly aluminum, titanium, and nickel.

The Daedalus design team, led by Alan Bond and Anthony Martin of Culham Laboratories, Oxfordshire (working in their spare time, not in an official capacity), decided from the outset that of the possible nuclear reactions which could power the vehicle, the fusion of deuterium and helium-3 was the most favorable, because it releases considerable energy but, unlike reactions between pure deuterium, or between deuterium and tritium, it does not yield neutrons, which would otherwise have to be screened from the rest of the craft by heavy shielding. However, subsidiary reactions in the starship exhaust do produce neutrons—enough that anyone within 400,000 kilometers would receive in excess of the safe human dose. Daedalus will therefore be taken a long way from Earth before its engine is ignited.

One problem is that there is not enough helium-3 on Earth to make the 30 billion tiny spherical bombs that would be needed. But there is sufficient helium-3 in the gaseous atmosphere of Jupiter to fuel millions of Daedalus starships. The Daedalus team foresee sifting the required 30,000 tons of helium-3 (a light isotope of helium) and 20,000 tons of deuterium (a heavy isotope of hydrogen) from the atmosphere of Jupiter using separation plants floated beneath balloons. Again, this would be within the capability of space industrialization in the next century, although it demonstrates that only a civilization with a solar-system-wide economy, like that which follows establishment of the O'Neill colonies, is likely to build and launch starships. A civilization begins interstellar travel once it has colonized its own solar system—by which time it will already be sending out messages to the stars (see Chapter 8).

Propellant bombs for the first stage of Daedalus are stored in six spherical tanks arranged in a ring around the vehicle's central core. The bombs are injected one at a time by a magnetic gun into the first-stage reaction chamber, a hemisphere 100 meters in diameter. Each bomb, a deuterium honeycomb filled with helium-3, is coated with a superconducting shell so that it can be shot into the reaction chamber by a magnetic wave traveling along a coil, which makes up the injec-

struments to survey the targets in wavelengths ranging from infrared to ultraviolet and to take detailed photographs. Special instrumented packages may be ejected to perform specific duties such as atmospheric entry and examination of satellites.

Daedalus carries two telescopes of 5-meter (200-inch) aperture, the same size as the Mount Palomar reflector, with which it begins to survey the Barnard's star system ten years before encounter. Such large telescopes are necessary because the planets would appear so faint in the weak light from their red-dwarf parent. Seven years before encounter, the first set of probes is dispatched on courses that will take them to intercept the largest, Jupiter-like planets of the star. As Daedalus gets closer its telescopes will be able to distinguish any Earth-like planets that exist around Barnard's star, and it will deploy probes to examine these planets, as well as Barnard's star itself, up to one year prior to encounter. Each sub-probe has a nuclear-electric propulsion system for final maneuvers.

Since Daedalus does not decelerate, it and its array of sub-probes will pass through the entire Barnard's star system in only a few hours; the encounter time of any one probe with a planet is only a few minutes. Yet during this brief period the Daedalus mission may be grabbing many thousands of photographs to give us our first close-up look at another planetary system in space. It will take several years to return this data to Earth.

Each sub-probe will dispatch its data to the mother ship, which is the top stage of the craft that left our solar system fifty years previously. The unwanted reaction chamber of the second-stage engine, which would long since have shut down and cooled, can be used as a giant radio telescope for communication with Earth. Down this radio link to large receivers on or near Earth will come the photographs and data from our first interstellar mission.

Although Daedalus has been designed with existing or attainable technology in mind, such a mission could still not come about overnight. Once the go-ahead had been given, detailed design, manufacture, and vehicle checkout with trial flights to the edge of the solar system and back would involve about twenty years, followed by fifty years' flight time and six to nine years for transmission of data to Earth. Therefore, the Daedalus team conclude, it appears that even the simplest interstellar missions will require a funding commitment lasting seventy-five to eighty years.

What would such a mission cost? One estimate from the Daedalus team is that it would cost something like $100 billion, or more than the United States has spent on the space program since the launching of its first satellite. Curiously, however, this happens to be exactly one year's defense spending by the United States. Therefore, if we forwent defense spending for one year, we could go to the stars.

In practice, of course, we could not attempt it today; but by the next century it would be possible, and I believe we will do it then. What better way to celebrate the 100th anniversary of the space age in 2057 than by launching our first ship to the stars?

Gregory Matloff of New York University's division of applied science has considered the possibility of attaching a Daedalus-type engine to an O'Neill-type colony. Such a combination ought to achieve a cruise velocity of one or two hundredths the speed of light, giving journey times of 400, 600, and 1,100 years to Alpha Centauri, Barnard's star, and Tau Ceti respectively.

As long ago as 1964, Robert D. Enzmann of the Raytheon Corporation proposed an interstellar ark driven by eight nuclear pulse rockets. The living quarters of the starship, habitable by 200 people but with room for growth, would be in three modules interlinked into a cylinder 300 meters long and 100 meters in diameter. Like the O'Neill colonies, this cylinder would spin to provide gravity. At the head of the starship would be a sphere 300 meters across filled with 12 million tons of frozen deuterium, which acts as the starship's fuel and also incidentally serves as a forward shield for the living quarters during the journey. Fuel is channeled from this giant snowball-on-a-stick through a central load-bearing core to the eight engines at the rear of the ship. Abundant neutrons are liberated in a deuterium-deuterium reaction, so that thick shielding will be needed around the engines to prevent danger to the crew.

Enzmann foresaw a fleet of three to ten such starships setting out together for their stellar destination, which they would reach at a speed of about one-tenth that of light. The modular construction of the vehicle means that even if one section is completely destroyed, it can be discarded while the other two carry on; in the event of a total catastrophe, the remaining crew can bail out into another member of the fleet, which will have spare room for them.

An adventurous fifty-year crash program leading to manned interstellar exploration has been proposed by Robert L. Forward of Hughes Research Laboratories in Malibu, California. On Forward's time scale, the first target would be to launch robot probes to the stars by the end of the century. The program would start with fifteen years of design and research into critical areas of technology, such as fusion drives. Funding during this stage would be limited to a modest few million dollars a year, but would climb to several billion dollars a year as the building and test flying of small prototype starships began, peaking around the year 2000 as the first full-scale probes are launched to Alpha Centauri, Barnard's star, and other likely nearby targets. Development work on manned versions of the craft would continue during the following twenty years while the probes, traveling at one-third the speed of light, reached their targets and sent back data. Once assured that all was well by the success of the robot pathfinders, manned missions would leave around the year 2025, reaching Alpha Centauri in ten to twenty years.

One must admire Forward's optimism, while admitting that his time scale is somewhat compressed. Perhaps if everything were stretched out four times in length, it might bear some relation to reality. Can we really consider flight to the stars before men have landed on Mars? Do we even want to send men to the stars when intelligent machines can do the job for us? These are questions that different societies might answer in different ways—and which we ourselves might view differently in decades to come.

One well-known starship design is the so-called interstellar ramjet, a device which in theory would scoop in fuel from the hydrogen clouds in space and burn this in a reactor to create thrust. The ramjet design, first suggested in 1960 by Robert Bussard, has the apparent advantage that it need carry no fuel and would gather in more hydrogen from space the farther and faster it went, making it seemingly ideal for long-distance, high-velocity voyages that would be necessary to reach distant parts of the Galaxy. But is it feasible? Bussard himself realized that such a design was beyond the reach of contemporary technology, but nuclear physicist Tom Heppenheimer of the Max Planck Institute in Heidelberg now claims that the interstellar ramjet would never work successfully, even in theory. It would be quicker to get out and push.

A ramjet would have to operate by fusion of deuterium (heavy hydrogen) from among the hydrogen clouds of space. Energy generation as in the Sun, by the fusion of protons (hydrogen nuclei), would release energy at a far slower rate—so slowly that in terms of power-to-weight ratio a galley full of Roman slaves would be more efficient; one could get more energy from the slaves' body heat while they were asleep, Heppenheimer notes puckishly. Even so, there are problems with the deuterium-deuterium reaction.

If the gas in the reaction chamber is not dense enough, it will radiate away more energy than it releases—at least 1 billion times as much as can be produced by the fusion reactions, according to Heppenheimer. The answer, it would seem, is to collect more gas and make it denser. This would require a ramscoop an astounding half-light-year in diameter. But even so, hydrogen is so thinly spread in space that more energy is required to compress it sufficiently than can be recovered from the subsequent fusion reactions. In either case, we lose more than we gain. Heppenheimer therefore concludes that the interstellar ramjet is infeasible for propulsion, but that it dissipates energy so successfully that it would make a very good brake—an effect that one day may prove useful, even if it is the reverse of the application that the ramjet advocates have had in mind.

By contrast, a device such as Daedalus carrying its own propellants emerges successfully from the analysis by Heppenheimer, who proposes that fusion engineers and starship designers should join forces to advance each other's art. A bomb-powered spacecraft may not seem the most elegant of engineering solutions, so perhaps things can be improved by further advances in fusion technology during the coming decades.

Looking into the realms of the fantastic, perhaps the ultimate form of propulsion one can envisage is the photon rocket, in which matter and antimatter annihilate each other to produce pure energy. This process is over a hundred times as efficient as hydrogen fusion, in which only a fraction of a percent of the mass is turned into energy, but it suffers from some astounding engineering problems, not least being to produce and store sufficient antimatter and then to generate thrust from the photons so produced. Even if these details can be resolved, not many starship captains would, I suspect, be happy with the thought of a craft containing a propellant that promised to annihilate any part of the ship it came into contact with. An accident in a starship whose antimatter fuel broke out of confinement would be marked by a flareup like the nova explosion of a star.

One intriguing possibility that I expect the science-fiction writers will soon be on to is to capture a mini–black hole for use as a starship power source. Work by theorists such as Stephen Hawking of Cambridge has shown that black holes can actually radiate energy, and that small black holes, containing the mass of a mountain squashed into a space the size of an atomic particle, radiate thousands of megawatts. Such mini–black holes could have been formed in the conditions of extreme temperature and pressure shortly after the Big Bang explosion believed to have marked the beginning of the Universe, and if it is possible to find and harness these compact energy sources, then one of the biggest problems of starflight will have been solved.

For the foreseeable future, though, we are stuck with the energy of nuclear fusion, as utilized by the Daedalus designers, which should be adequate for our purposes. The importance of the Daedalus study, the first detailed design for an interstellar mission, is that even if it bears no relation to starships as eventually built, it will still have served its purpose by showing how we *might* build our first ship to the stars. Although we are, for the moment, restricted to making plans on paper, other civilizations more advanced than ourselves may have already put those plans into practice. If there are starships of other civilizations in space, could we expect to see them?

The possible detection of starships has been discussed by David Viewing of the British Interplanetary Society, with colleagues C. J. Horswell and E. W. Palmer. The authors believe that slow, unpowered starships like the Pioneer and Voyager craft will be the most abundant at first, because they are so much cheaper than the high-powered Daedalus kind and they will eventually return just as much data if we are willing to wait long enough. Sufficiently reliable electronic systems to last the journey should be feasible. Probes like this would be virtually undetectable unless someone possessed an intensely powerful interstellar early-warning radar for spotting stray bodies in space. Beams from such a radar should be detectable on Earth, although radio astronomers have to date found no signs of any artificial interstellar transmissions. But what about larger probes with high-energy propulsion systems, like the fusion drive of Daedalus?

Daedalus itself is not designed to have its engine firing as it flies past its target star. It would be invisible by reflected sunlight as it passed through our solar system even on photographs taken with large telescopes, because it is too small and is moving too quickly. A large enough craft might reveal itself by briefly blocking out light as it passed in front of stars, but the chances of detection in this way are extremely slim. Viewing and his colleagues examined the possibility of a net of sentry probes around the Sun designed to detect itinerant starships, and found that a ring of approximately 1,000 probes would be needed to be sure of spotting any alien Daedalus-like craft that flew between them.

"Daedalus probes could arrive at our system at hourly intervals completely unnoticed by terrestrial civilization," conclude the authors. "No matter how awesome the starship might be in a terrestrial context, in its own environment—interstellar space—it is virtually invisible."

Starships with more advanced capabilities than Daedalus hold out more hope of detection, because the emissions from their engines would give them away. "It seems probable that the radiation and particle fluxes generated in the engines of starships are, even at this moment, falling upon the Earth," say Viewing and his colleagues. The problem is to separate this influx from the natural cosmic-ray background. "The fires of starship engines are lost at present in galactic noise."

A starship would be easily distinguished from a normal cosmic-ray source because of its high-velocity motion, which may also allow us to trace back its path to the star of origin. A starship's engine will also be a point source, far smaller than any natural galactic source, and in the case of a Daedalus-type engine the emission will be rapidly pulsed. Although a starship's engine might be shut down for the coast between stars, it would light up again when maneuvering into or out of a solar system. We might hope to spot a starship maneuvering in our own solar system.

Fusion explosions in a Daedalus-type starship engine give out X rays, and neutrons are produced from subsidiary reactions in the exhaust. From these emissions, the second stage of Daedalus could be detected by instruments like those in satellites and on Earth today at distances of 60 to 70 million kilometers; in other words, from Earth it could be detected beyond the orbit of Venus almost out to the orbit of Mars. Daedalus's exhaust consists of highly ionized hydrogen, deuterium, and helium, which would produce a glow of light in the far ultraviolet that could be seen by suitable instruments at more than 1 billion kilometers, the distance of Saturn.

Viewing and his colleagues note that throughout Earth's history, less advanced civilizations have usually first detected the more advanced varieties by observation of their vehicles—first sailing ships, and now aircraft. Even the most remote tribe on Earth could not by now have failed to see an aircraft or a satellite. We, so far, have failed to detect even one interstellar spacecraft, although we could certainly expect to have done so had there been any maneuvering among the inner planets of the solar system. The fact that we have made no such observations of starships near the Earth is one argument against the proposition that we are currently being visited by extraterrestrial space travelers. Other objections are given in Chapter 16.

REFERENCES

Forward, R. L., *Journal of the British Interplanetary Society*, Vol. 29 (1976), p. 611.

Matloff, G. L., *Journal of the British Interplanetary Society*, Vol. 29 (1976), p. 775.

Project Daedalus Study Group, *Spaceflight*, Vol. 19 (1977), p. 419.

Project Daedalus Final Report, *Journal of the British Interplanetary Society*, Supplement, 1978.

Schroeder, T. R., *Astronomy*, August 1977, p. 6.

Viewing, D. R. J., C. J. Horswell, and E. W. Palmer, *Journal of the British Interplanetary Society*, Vol. 30 (1977), p. 99.

7

Messages from the Stars

Eventually we will send probes to the stars in search of life, but at the moment our space-probe exploration is confined to our own solar system. Even if there is life in the solar system—and in the wake of Viking that seems improbable—it will be only of the most lowly kind. If we want to find signs of advanced technological life, like ourselves or more advanced, we must listen for messages from the stars. Several pilot listening programs have been undertaken by individual radio astronomers, and now both NASA and the Soviet Academy of Sciences are embarking on ambitious signal searches.

In January 1900 the radio pioneer Nikola Tesla described strange signals he had received on his own radio equipment the previous year: "The changes I noted were taking place periodically, and with such a clear suggestion of number and order that they were not traceable to any cause then known to me . . . It was some time afterward when the thought flashed upon my mind that the disturbances I had observed might be due to intelligent control." This was the first suggestion that radio communication between worlds in space might be taking place. Bearing in mind the low sensitivity and the long operating wavelength of Tesla's equipment, it seems unlikely that he could have been detecting an interstellar message, but the exact nature of the disturbances he recorded was never clarified because he refused to give further details after his speculations about interstellar radio communication had been harshly criticized.*

Nowadays, the idea of interstellar radio communication has become quite widely accepted. With modern radio telescopes we could make our presence known by signaling to nearby stars, so naturally the question arises: Are other civilizations trying to contact us? We could easily pick up their transmissions with existing radio telescopes, if any such transmissions exist. But there are two main problems: Where to look? And at which wavelength?

Imagine not being able to pick up a radio station on a domestic receiver unless the aerial were pointed directly at the transmitter, and having to tune through all the bands to find the right channel. That's the problem faced by radio astronomers, with the added disadvantage that it may take several minutes' listening at each wavelength to pick out the faint signal from the background noise. Fortunately, we can make a few assumptions that simplify the problem.

As noted in Chapter 2, astronomers believe that stars similar to the Sun are the most likely to harbor advanced extraterrestrial life because they are hot enough to warm any planets that may orbit them (unlike smaller, fainter stars) and they keep up their stable light output long enough for advanced life forms to develop (unlike larger, hotter stars which burn out too quickly). Therefore, the signal searchers have concentrated on Sun-like stars, plus some nearby red dwarfs like Barnard's star, with the emphasis on single stars because of the doubts about the existence of planets in double-star systems.

The first serious attempt at detecting messages from the stars was Frank Drake's now classic Project Ozma in 1960, during which he listened fruitlessly for 150 hours to the Sun-like stars Tau Ceti and Epsilon Eridani with the 26-meter dish at the National Radio Astronomy Observatory at Green Bank, West Virginia. Drake listened at a wavelength of 21 centimeters, which is naturally emitted by hydrogen in space (21-cm radiation is equivalent to a frequency of 1,420 megahertz). Since hydrogen is the most abundant substance in the Universe, its 21-cm line will be known to radio astronomers everywhere who will be regularly tuned to it, and so it would make a good band for interstellar signaling.

* Tesla's "extraterrestrial signal" is reported by Leland I. Anderson in *Nature*, Vol. 190 (1961), p. 374.

The idea of listening for messages from the stars at 21-cm wavelength, first suggested in 1959 by the physicists Giuseppe Cocconi and Philip Morrison, has been widely adopted by subsequent signal searchers. In 1972 radio astronomer Gerrit Verschuur at the National Radio Astronomy Observatory updated and extended the Project Ozma search by listening at 21-cm wavelength with the Green Bank 43-meter and 91-meter telescopes to ten nearby stars, although again with negative results.

The most comprehensive search undertaken to date was completed in 1976 at Green Bank by Benjamin Zuckerman and Patrick Palmer. During the previous four years they had surveyed at 21-cm wavelength the 659 stars most likely to harbor life from 6 to 76 light-years from the Sun, using the Green Bank 91-meter and 43-meter dishes. Zuckerman and Palmer estimated that with the vastly increased sensitivity of their equipment over the original Ozma experiment, they could have detected a 40-Mw transmitter beaming through a 100-meter dish (a technical capability equivalent to our own) around any of the target stars. They listened to each star for four minutes at a time, reobserving them six or seven times in all.

Ten stars which had shown "glitches" (i.e., unexplained spikes of energy) were carefully resurveyed. Several of the glitches were traced to terrestrial sources of interference, such as aircraft, while others remained a mystery; however, since none of the spikes were repeated, they were unlikely to be due to beacon transmissions from other civilizations.

Another extensive search, still under way, is being undertaken by Robert S. Dixon at the Ohio State University radio observatory. Dixon is concentrating on the 21-cm hydrogen line, but with a slight difference. Because all stars are moving in orbits around the Galaxy, their motions cause a slight wavelength change, known as the Doppler shift, in any signals transmitted or received from them. Therefore, a signal transmitted at 21-cm wavelength by another civilization would be received at a slightly different wavelength, depending on whether we were moving toward or away from them. The only way to avoid this problem, thinks Dixon, is if both the receiving and the transmitting civilizations correct for their own motion around the galactic center so that any signal would be received at a constant wavelength, as though it were being transmitted from a stationary aerial at the center of the Galaxy.

Rather than concentrate on specific targets, Dixon is scanning the entire sky from 63°N to 36°S visible with the radio telescope originally used to compile the Ohio Sky Survey, a famous radio astronomy catalog. This telescope consists of an antenna 100 meters long and 30 meters high, whose collecting area is equivalent to that of a 53-meter circular dish. In December 1973, Dixon began his search with his detector centered on the frequency that would be received from a 21-cm transmitter at the center of the Galaxy, running 24 hours a day, 365 days a year. He estimates that out to a distance of 1,000 light-years his telescope beam covers an average of three stars like the Sun, or slightly hotter or cooler than the Sun, at all times. Such an all-sky search has the added advantage that it would detect messages coming from the space between stars, such as might be transmitted from an interstellar probe or a space colony.

By the end of 1977, Dixon and his assistant Dennis M. Cole had covered the northern sky from declinations +48° to +14° and the southern sky from declinations −23° to −33° without detecting any apparently artificial signals. Dixon plans to continue his search indefinitely, continually improving the sensitivity of his receiver to pick up ever fainter signals. By the time the first search is completed, the improvements in sensitivity will make it worthwhile to begin another search.

Frank Drake and Carl Sagan of Cornell University have used the largest single radio dish in the world, the 305-meter dish at Arecibo, Puerto Rico, to listen for messages coming from possible supercivilizations in nearby galaxies. Although a civilization so advanced as to be able to make itself heard over intergalactic, rather than interstellar, distances would be rare, the advantage of looking at a complete galaxy is that it brings billions of stars into view at once. Beginning in 1975, Drake and Sagan scanned four nearby galaxies at wavelengths of 21, 18, and 12.6 cm without detecting any suspicious signals.

Their choice of reception wavelengths was governed by the equipment available at the radio telescope, a common limitation in radio astronomy. The 12.6-cm wavelength is used by the Arecibo dish for interplanetary radar bounces, but the 18-cm line is more interesting. It is the wavelength emitted by the molecular grouping known as hydroxyl, or OH, which is water lacking one hydrogen atom. The 18-cm and 21-cm wavelengths of OH and H stand like gateposts on either side of an area of the radio spectrum commonly termed the water hole. Many radio astronomers are beginning to regard the water hole as the preferable slot for interstellar transmissions, for two reasons: because it coincides with a natural minimum in the background noise from space, and because of the importance of water to all life as we know it. In recognition of this, Robert S. Dixon has now incorporated a receiver to scan the area around the 18-cm OH line in addition to his 21-cm search at the Ohio State University radio observatory.

There are certain limits on the observable radio window. At wavelengths longer than about 30 cm, background noise from our Galaxy becomes a problem. At wavelengths shorter than about 3 cm, noise from our atmosphere, particularly water vapor, becomes obtrusive. Water emits strong signals at a wavelength of 1.35 cm, and thus defines another possible line for interstellar communication, which, to some astronomers, seems as appealing as the 21-cm line. Two Canadian radio astronomers, Paul Feldman and Alan Bridle, began a search for possible interstellar messages at the 1.35-cm water wavelength in 1974, using the 46-meter dish at the Algonquin radio observatory in Ontario. Their plan is to examine around 500 likely stars, but the project has been severely hampered by equipment failures and has so far not progressed beyond surveying about 70 stars.

Some scientists involved in the search for extraterrestrial intelligence (SETI*) do not even think that radio is the most likely mode of communication. An American engineer, Herbert Wischnia, has argued that lasers sending beams of ultraviolet light are a likely means of interstellar communication. Since stars like the Sun radiate relatively little energy in the ultraviolet, the laser beam would not be swamped by natural stellar radiation. Ultraviolet light is absorbed by the Earth's atmosphere, so the only way to check this speculation is by satellite, and Wischnia obtained observing time in late 1974 on NASA's OAO-3 satellite, also known as Copernicus, which has an 81-cm telescope on board for examining ultraviolet radiation from stars. With this observatory satellite, Wischnia looked at Epsilon Eridani, Tau Ceti, and Epsilon Indi, but found no ultraviolet laser emissions.

Cambridge astronomer Andy Fabian has made the somewhat

* NASA now prefers this acronym to the previous CETI (communication with extraterrestrial intelligence) because of the erroneous impression of a two-way link given by the word "communication."

tongue-in-cheek suggestion that advanced civilizations might attempt to attract our attention by dropping matter onto neutron stars to produce bursts of X rays. A 1-km chunk of rock dropped onto a neutron star would produce a flash of X rays detectable throughout the Galaxy. However, there is no indication that any of the X-ray sources observed by satellites are actually interstellar signals.

Some civilizations might go so far as to vary the light output of a star, like a galactic Aldis lamp. Many stars are now being discovered to exhibit fractional fluctuations in light output, detectable only by sensitive instruments. One suggestion is that these light changes are caused by slight oscillations in size of the star. An advanced civilization could set an entire star "ringing" to produce brightness fluctuations by impacting material such as comets onto the star's surface; they might even modulate a coherent message on the fluctuations.

While most astronomers strongly favor radio communication as the best way of establishing contact between stars, the idea of using the 21-cm hydrogen line has come under increasing attack, because any signal would tend to be swamped by the background noise of natural hydrogen radiation, and also because a powerful 21-cm transmitter would jam all local receivers. This second disadvantage, at least, is avoided by Robert S. Dixon's strategy of applying a frequency correction in all directions at the transmitter and receiver to compensate for the parent star's rotation around the Galaxy.

Some astronomers have proposed using harmonics, or multiples, of the 21-cm line, but these could be mistaken for natural emissions. A more intriguing idea has been proposed by P. V. Makovetskii of the Leningrad Institute of Aviation Instrument Manufacture. He agrees with Cocconi and Morrison that the 21-cm line of hydrogen will be universally known to astronomers, but he also notes that certain mathematical constants, such as π and $\sqrt{2}$, should also be universally recognized. His proposal for a clearly recognizable artificial signal is to beam radio transmissions at a specific wavelength defined by a combination of these physical and mathematical constants. He regards the six most probable wavelengths as being the 21-cm hydrogen line alternately multiplied and divided by π, 2π, and $\sqrt{2}$. The resultant wavelengths, 3.3 cm, 6.7 cm, 15 cm, 30 cm, 66 cm, and 132 cm, will be precisely defined, and would be immediately identified as artificial.

A precisely defined wavelength is an advantage, because the

narrower the bandwidth of a signal (that is, the smaller the spread of frequencies) the more powerful the signal will be for a given transmitter output, and therefore the greater its range. The disadvantage, from the point of view of radio astronomers, is that their receivers are tuned to accept a fairly wide band of frequencies, usually several kilohertz, which reduces their sensitivity to narrow-band signals; the wider a receiving channel, the more difficult it is to pick out a signal from the noise from surrounding frequencies, which is perhaps one reason why messages from the stars have not yet been detected during normal radio astronomy observations. And conversely, the narrower a receiver channel, the more time it takes to scan a given region of the spectrum, such as the water hole, or the more receivers operating in parallel are needed to cover the same frequency range simultaneously.

Another "fundamental" frequency has been proposed by radio astronomers T. B. H. Kuiper and M. Morris, based on the speed of light and certain properties of the electron; the resulting frequency is equivalent to a wavelength of 11.7 cm. Kuiper and Morris have also discussed possible search strategies, arguing that messages would most probably be coming from nearby rather than far off, if they are coming to us at all. The reasoning behind this claim is that if there are many civilizations out there which are millions of years in advance of us, they will most likely have already colonized the Galaxy (see also Chapter 11). We might therefore come under the aegis of one local area of this galactic empire, whose role would be to attract our attention when they thought we were ready. Therefore, argue Kuiper and Morris, if we are being signaled to at all, the message is probably coming from no farther away than 10 light-years or so.

In the wake of the Zuckerman-Palmer and Dixon searches, new SETI projects have been springing up in the United States. Radio astronomers at the University of California at Berkeley have begun a SETI program with the 26-meter dish at the Hat Creek radio observatory. The search is not aimed at specific targets, but runs in conjunction with normal radio astronomy work being done at the observatory. The wavelengths under scrutiny are 21 cm, 16 cm, and 32 cm.

In 1976 and 1977, a combined team from NASA's Ames Research Center and Goddard Space Flight Center used the Green Bank 43-meter and 91-meter telescopes for narrow-band examination of 200 stars, including those which showed glitches in the Zuckerman and

Palmer search, at wavelengths of 18 cm and 35 cm. Their receiver, which has a bandwidth of 5 Hz, better than anything used previously for SETI, was designed for particularly high-resolution radio astronomy work. At Arecibo, Frank Drake has also begun examination of several likely stars at 18-cm wavelength with a receiver of even narrower bandwidth.

These latter two narrow-band searches are predecessors of even more ambitious plans officially backed by NASA, and announced in 1977. On one thing the NASA searchers seem agreed: the incoming signals are likely to be of very narrow bandwidth, so that special receivers are required. However, since there now seems no reason to choose any particular wavelength, the searchers are intending to scan as much of the radio spectrum as possible, particularly in the water hole.

To speed up the search, which would otherwise be prohibitively long, NASA's Ames Research Center and Jet Propulsion Laboratory, both in California, are cooperating on the development of a device known as a Multi-Channel Spectral Analyzer (MCSA) which will be able to survey a wide range of radio frequencies simultaneously. The MCSA under design will have one million channels, so that it can survey a 300-MHz band (the width of the water hole) with a resolution of 300 Hz per channel. Alternatively, it can be used to survey a smaller slice of the radio spectrum with resolutions of only a few Hertz per channel. Eventually there might come a 1 billion-channel MCSA, capable of scanning the entire water hole at a time with 0.3-Hz resolution. Such a staggering increase in capability over existing receivers is due largely to vast improvements in microelectronics, which can now cram such a powerful package into a manageably small size.

The MCSA will feature in a five-year SETI plan being embarked upon jointly by the Ames Research Center and the Jet Propulsion Laboratory. Both the Ames and the JPL searches will have greater sensitivity than their predecessors. The Ames team will be using the Arecibo dish and two other radio telescopes, one in the Northern Hemisphere and one in the Southern, to scan the water hole in the direction of at least several hundred selected stars within 1,000 light years, with resolutions of a few Hertz. According to astronomer David Black of the Ames team, this program will take only a few percent of the observing time on each telescope, and has an anticipated cost of $6 million, most of which is accounted for by the MCSA and equipment to analyze the signals for possible message content.

The JPL team are making even fewer assumptions, intending to scan the entire sky between 30-cm and 1.2-cm wavelength with 300-Hz resolution, using an identical MCSA attached to the Goldstone 26-m tracking dish, plus a network of smaller horn antennas. According to the manager of the JPL project, Robert E. Edelson, the Goldstone dish will be able to cover about 80 percent of the sky. The 26-m dish is used about 40 hours a week for deep-space tracking, but for the rest of the time, about 16 hours a day, it will be available for the automated SETI program. The cost of the JPL program is estimated to be $14 million, bringing the combined outlay for both searches to $20 million—equivalent to the cost of the movie *Close Encounters of the Third Kind.* Subject to budgetary approval, building of the MCSAs was due to begin in October 1978, with field tests in mid 1979. On this schedule, the five-year program of SETI observations will begin in 1980.

Even if neither project detects messages from the stars, the surveys will produce considerable new astronomical information so that the effort will not be wasted. Only such a coordinated program, rather than the piecemeal attempts made to date, will provide firm evidence about the existence of communicative civilizations elsewhere in space. "Within a few decades, mankind will either have discovered the presence of extraterrestrial intelligence, or will have placed very severe constraints on the likelihood that such intelligence exists," says David Black.

American astronomers are not alone in their SETI efforts. Soviet radio astronomers have been making attempts to detect messages from the stars since 1968. Unlike their American counterparts, the Soviet astronomers have tended to use modest receivers in the hope of detecting highly powerful transmissions from super-civilizations, with the receivers spread out across the USSR to pinpoint the source of any strange signals received. In 1973 the group led by Dr. Nikolai Kardashev was reported to have picked up coded signals which they believed might be from extraterrestrials. But, according to a 1977 report by the Soviet science journalist Boris Belitsky, subsequent work showed the signals to have come from Earth satellites.

Although these initial Soviet attempts were modest, an altogether more ambitious ten- to fifteen-year SETI program was approved in 1974 by the Soviet Academy of Sciences. The SETI program is split into two parts: SETI 1, from 1975 to 1985, and an overlapping SETI 2 from 1980 to 1990. The proposals for SETI 1 include all-sky monitoring from eight ground stations, with associated antennae for continuously surveying nearby galaxies; this is an extension of existing Soviet SETI work. Additionally, two space stations with large antennae are visualized to augment the all-sky search. Features of SETI 2 are enlargement of the space station antennae and introduction of a widely spaced pair of dishes with a collecting area of 1 square kilometer for examining specific objects.

The SETI program recommends scanning radio and infrared frequencies from 0.3 cm to 30 cm, but with special emphasis on molecular lines such as 21 cm, 18 cm, and 1.35 cm. The new RATAN 600 radio telescope, consisting of a group of 900 collector plates arranged in a ring 600 meters in diameter, is planned to cover the wavelengths from 21 cm to 0.8 cm, while a millimeter-wave telescope is under development at Gorki for even shorter-wavelength studies. In addition to searches of globular clusters and the dense star fields at the center of the Galaxy (both being locations of old stars where we might find highly advanced civilizations), and galaxies of the local group, the Soviet SETI plans recommend monitoring "all appropriate stars" up to 100 light-years distance, and eventually out to 1,000 light-years. In the future, Soviet researchers foresee automated probes being sent to investigate likely targets.

It's difficult to know how much of the program will actually be achieved, as it seems more of a statement of desirable aims rather than an instruction to be followed to the letter. But it does at least seem clear that Soviet scientists are taking SETI seriously, and that it is becoming an adjunct of normal astronomical investigation.

What of the future? If the existing searches fail to detect signs of messages from the stars, we may have to consider new approaches to the problem; for example, casting our net as widely as possible and attempting to monitor the whole sky continuously. To assist with all-sky monitoring, Hendrik J. Gerritsen and Sean J. McKenna of the physics and astronomy department at Brown University have proposed using a device known as the Luneberg lens, named after R. K. Luneberg, who invented it in 1944; with this device, say Gerritsen and McKenna, we can simultaneously monitor "a huge number of stars." A Luneberg lens is a large sphere filled with a suitable medium which refracts radio waves onto its inside back surface, thus forming an image of the radio sky. One Luneberg lens can image a hemisphere; two could cover the entire sky. The Luneberg lens is the receiving equivalent of an omnidirectional beacon.

Signals from each of the many star images focused on its periphery are collected by an equally large number of microwave horns and guided to an amplifier. Experimental spheres made of lightweight polystyrene foam have been constructed. According to Gerritsen and McKenna, a Luneberg lens of a given diameter performs as well, if not better, than a dish reflector of the same aperture up to 70 meters across, at which point performance of the Luneberg lens falls off due to losses from absorption of microwaves in the sphere.

Gerritsen and McKenna conclude that a Luneberg lens would be best placed on the Moon or in space for a SETI program which they term Project Argus. Assuming that other civilizations are already listening with a Luneberg lens, they believe we should begin a program of short transmissions to the nearest stars, and then begin listening for replies. At the moment, though, the idea of the Luneberg lens for SETI seems to have met with little response.

NASA has studied the possibility of building several advanced systems for detecting messages from the stars, either on the Earth, the Moon, or in space; such systems will be necessary if signals turn out to be extremely faint, or distant, or both. Most famous of these is the Cyclops project, an array of 1,000 or more dishes each of 100-meter aperture; together these would equal the performance of a single dish 3 kilometers in diameter, sufficient to detect a 1,000-Mw omnidirectional beacon with 0.1-Hz bandwidth at distances of 500 light-years, an area that includes several hundred thousand stars of interest. Transmissions of only a few megawatts strength could be detected from much closer stars, such as Tau Ceti and Epsilon Eridani, including leakage from domestic TV or radio broadcasts not even intended as interstellar signals.

A study group at the Stanford Research Institute, acting on a NASA contract, has compared an Earth-based Cyclops with other systems in space or on the Moon.* A lunar system has the advantage that, on the Moon's far side, it would be shielded from the radio noise of Earth; the lower lunar gravity is also an advantage when building large dishes. But, of course, costs would be enormous. The Stanford team looked at the possibilities of building a lunar Cyclops with steerable dishes, and even a lunar Arecibo. Whereas the Arecibo dish on Earth is suspended in a natural hollow between mountains, the Stanford team envisaged hundreds of similar dishes suspended from the rims of small bowl-shaped lunar craters. A comparable multi-dish Arecibo array on Earth would be hampered by lack of suitable mountain hollows in which to hang the reflectors.

Although an array of Arecibo-type reflectors on the Moon works out at about half the cost of a lunar Cyclops, the dishes are unsteerable and can only sweep a limited range of sky, so that they would require a larger collecting area and would have to search out to greater distances to have the same chance of detecting a message. It would be far cheaper to build and operate large radio astronomy arrays on the Moon once lunar industrial bases are established, such as for building the O'Neill colonies (Chapter 5). In that case, the cost of a lunar Arecibo array would be similar to that of an Earth-based Cyclops.

The Stanford group also looked at large space antennae, possibly to be situated at the L3 Lagrangian point of the Earth-Moon system, which is on the opposite side of the Earth from the Moon. But any of the Lagrangian points, including the L4 and L5 positions suggested for the O'Neill colonies, could be used. The "cup" antenna would be backed by a "saucer" to shield it from terrestrial radio noise. Because of their potential for complete sky coverage, their ability to track a source continually should a message be discovered, and their simplicity of construction, space systems are immediately attractive.

Collectors in space with areas of up to a few square kilometers, capable of probing several hundred light-years, would cost two or three times as much as an equivalent Earth-based system. But one surprise from the Stanford study was that a space system might be the cheapest of all at the very large collector sizes, 5 kilometers or so in diameter,

* Parametric Study of Interstellar Search Systems by Roy P. Basler, George L. Johnson, and Richard R. Vondrak, Stanford Research Institute Project 4359, August 1976. See also Radio Science, Vol. 12 (1977), p. 845.

that would be needed to push the search for a 1,000-Mw omnidirectional beacon out to 1,000 light-years. Costs would be cheaper still following establishment of an O'Neill-type space manufacturing facility, which could turn out antennae of 2.5-km diameter for just over $200 million each, according to O'Neill's estimates.

The Stanford study concluded that Moon-based systems are the least cost-effective of all three options, but that the cost-effectiveness of Earth-based Cyclops or space systems depended on the number of civilizations in the Galaxy and thus the size of the collectors needed to detect messages from them. Overall costs are difficult to assess accurately, but the total Cyclops system is likely to cost at least $10 billion, or a substantial proportion of the cost of the Apollo Moon project.

A NASA committee on the search for extraterrestrial intelligence, chaired by Frank Drake, met in May 1976 to discuss the relative merits of these advanced proposals, without coming to any firm conclusion. In any case, even if such systems were approved, construction would be unlikely to begin until the 1990s. The SETI committee did, however, focus attention on the increasing use by terrestrial radars and satellites of the microwave radio region, particularly around the water hole. If such interference continues to grow, there will be no alternative but a space- or Moon-based SETI system.

But behind all this technological wizardry, what exactly is it that they hope to hear? What would a message from the stars sound like? According to communications specialists, the most efficient way to transmit information is to spread it out as widely in frequency as possible. Anyone overhearing such a transmission would not understand it unless he knew the code to extract the message that was being carried; otherwise the signal would sound like random noise. But if someone were deliberately trying to attract attention, he would presumably send a signal that was unmistakable—one confined to as narrow a bandwidth as possible, say 1 Hz of frequency or even less. And this obviously artificial narrow-band transmission would be the carrier for a message that would clearly be encoded on it, either as pulses or in the form of alternate left- and right-hand polarization.

What would this message contain? It could be very complex, perhaps divided into three overlapping parts. One part might be a regularly repeated introductory passage, perhaps with a translation guide to assist comprehension of the rest. In between this could be a longer and continually updated message containing current information on the planet, its population, and technological level—a kind of travelogue-cum-documentary. The third part of the message might be a continuing story of the history of the planet and the development of its people—a celestial slide show of anthropology, art, and culture.

Or the signal might be the simplest one capable of unambiguously announcing the transmitting civilization's presence—a steady narrow-band beacon. To get a better idea of the type of transmissions that we might expect to receive, let's take a look at our own first attempts at interstellar signaling, and what we might do in the future.

REFERENCES

Belitsky, B., *Spaceflight*, Vol. 19 (1977), p. 193.

Black, D. C., et al., *Mercury*, July–August 1977, p. 3.

Dixon, R. S., and D. M. Cole, *Icarus*, Vol. 30 (1977), p. 267.

Edelson, R. E. *Mercury*, July–August 1977, p. 8.

Fabian, A., *Journal of the British Interplanetary Society*, Vol. 30 (1977), p. 112.

Gerritsen, H. J., and S. J. McKenna, *Icarus*, Vol. 26 (1975), p. 250.

Kuiper, T. B. H., and M. Morris, *Science*, Vol. 196 (1977), p. 616.

Makovetskii, P. V., *Soviet Astronomy*, Vol. 20 (1976), p. 123.

Murray, B. C., S. Gulkis, and R. E. Edelson, *Science*, Vol. 199 (1978), p. 485.

Ridpath, Ian, *New Scientist*, Vol. 74 (1977), p. 326.

Sagan, C., and F. D. Drake, *Scientific American*, Vol. 232 (1975), p. 80.

Sheaffer, R., *Spaceflight*, Vol. 17 (1975), p. 421.

——— *Spaceflight*, Vol. 18 (1976), p. 343.

——— *Spaceflight*, Vol. 19 (1977), p. 307.

Soviet SETI Program, *Soviet Astronomy*, Vol. 18 (1975), p. 669 (reprinted in *Icarus*, Vol. 26 [1975], p. 377).

Verschuur, G., *Icarus*, Vol. 19 (1973), p. 329.

8

How We Could Talk to the Stars

At 1700 Greenwich Mean Time on November 16, 1974, mankind's first deliberate radio message to the stars was transmitted from the 305-meter-diameter radio telescope at Arecibo, Puerto Rico. It was a three-minute signal, sent just once toward a group of stars 24,000 light-years away—hardly a serious attempt at interstellar communication. But it was a demonstration of what might be possible on a wider scale in the future, as well as a pointer to what we might discover coming from other civilizations in the Galaxy.

The Arecibo message was transmitted as part of the reinauguration of the radio telescope after it had been resurfaced and new equipment installed. The Arecibo telescope's maximum power in any one direction is now equivalent to that of an omnidirectional antenna blasting out 20 million megawatts in all directions, ten times the generating capacity of all electric power stations on Earth. This upgrading of the instrument means that with a bandwidth of 1 Hz or less its signal can be detected by similar radio telescopes throughout the Milky Way. As the first use of this new facility, the staff thought it "highly appropriate" to send a simple and brief signal to the fringes of the Galaxy announcing our existence.

The Arecibo message was concocted by staff members of the National Astronomy and Ionosphere Center, headquartered at Cornell

University, which operates the radio telescope; primarily concerned were Frank Drake, Richard Isaacman, Linda May, and James C. G. Walker, with assistance from others, notably Carl Sagan. Their message consists of a string of 1,679 pulses, which can be arranged into a grid of 23 characters by 73, forming a dot picture that reveals details about humankind which might be of interest to extraterrestrials. The idea of sending streams of on-off pulses which can be arranged to give a picture is widely favored for interstellar communication, because of its essential simplicity and because picture communication overcomes all language barriers. Pulse trains can also convey complex mathematical concepts: computers use the same kind of on-off pulse system, known as the binary code. Many scientists involved in the search for messages from the stars expect that beacon signals from extraterrestrials will be pulsed.

Once the pulses of the Arecibo message are correctly assembled, they portray the following: a counting system, five biologically significant atoms (hydrogen, carbon, oxygen, nitrogen, and phosphorus), the component parts of DNA, a schematic view of the DNA double helix, and representations of a human being, the solar system, and the Arecibo dish itself.

The target for the message was a globular cluster of about 300,000 stars called M 13, in the constellation Hercules. Since the cluster is about 24,000 light-years away, no rapid reply is expected, even if anyone happens to pick up the transmission. M 13 is just the right size to fill the Arecibo telescope's beam, so that all 300,000 stars were covered simultaneously. The signal was transmitted at the rate of 10 pulses per second on 12.6-cm wavelength (2,380 MHz), with a bandwidth of 10 Hz. The chosen wavelength is used by astronomers for radar studies of the solar system, and is not one of the wavelengths particularly favored for interstellar communication. However, as noted in the previous chapter, signal searchers are now abandoning the idea of special interstellar communications wavelengths and are instead attempting to scan as much of the radio spectrum as possible; presumably aliens listening out for us would also be covering all options in the same way.

The Arecibo message is believed to have been the strongest signal yet radiated by mankind, equivalent to an omnidirectional beacon of 3 million megawatts. In the words of the Arecibo staff, it was intended as "a concrete demonstration that terrestrial radio astronomy has now reached a level of advance entirely adequate for interstellar radio communication over immense distances." They added, though, that more extensive attempts at interstellar signaling should not be made until after international consultation, as was agreed by representatives from the United States, the USSR, and other countries at an international conference on extraterrestrial communications held in Byurakan, Soviet Armenia, in September 1971. Even so, there were criticisms that the Arecibo message had been sent without prior warning to the signatories of the Byurakan agreement.

The Arecibo signal was not the first message for other civilizations to leave Earth. In March 1972 and April 1973 two spacecraft, Pioneer 10 and Pioneer 11, were launched on deep-space missions to the planet Jupiter, after passing which they would drift out of the solar system and toward the stars. Pioneer 11 was later retargeted to look also at Saturn, which it was scheduled to reach in 1979, but eventually it too will follow Pioneer 10 out into the Galaxy. Of course, both probes will have long stopped transmitting before they reach any stars; even if they were aimed at the nearest star, and they are not, they would take 80,000 years to get there. (The Arecibo radio message, traveling at the speed of light, overtook the Pioneer probes about an hour after it was transmitted.) But on the off chance that the probes may one day swim into the ken of some advanced galactic civilization which will retrieve them, both Pioneers carry identical engraved plaques telling something of their place of origin and the people who sent them.

One part of the Pioneer plaque shows the position of the Sun relative to fourteen pulsars. Since any civilization advanced enough to retrieve a space probe will know about pulsars, they will be able to reconstruct the approximate position of the Sun in the Galaxy. The periods of the pulsars are indicated in binary notation; since pulsars run down with age, the receiving civilization will also be able to work out roughly when the probes were sent by comparing the observed frequency of each pulsar with the frequency indicated on the plaque. Another part of the engraving shows the solar system, with the path of the Pioneers past Jupiter.

But most puzzling of all may be the representations of human beings. What will other creatures make of these? Much debate has been occasioned by the fact that the man has one hand raised in what is alleged to be the universal gesture of greeting and friendship. But it would be foolish to imagine that it would be interpreted so everywhere in the Galaxy; to some other civilization it could be extremely rude. I once encountered a cageful of Rhesus monkeys, who are not too far removed on the family tree from man, and I decided to try them out: I raised my hand to the side of their cage in the universal gesture of greeting and friendship. And they attacked me.

In August and September 1977, the Pioneer probes were joined in space by two probes carrying messages of a different kind: sounds of the Earth on a 12-inch long-playing record. The probes were Voyagers 1 and 2, bound initially for the outer planets but eventually destined, like the Pioneers before them, to leave the solar system and drift forever among the stars.

As with the Pioneer plaques, the man behind the scheme was galactic guru Carl Sagan of Cornell University. His colleagues involved were Frank Drake, Alastair Cameron, Philip Morrison, Bernard Oliver, and Leslie Orgel, in addition to musicians and artists. A record was chosen because it can contain more information than a simple engraved plaque, and also to commemorate the 100th anniversary of Edison's invention of the phonograph.

Each record, made of copper and placed in an aluminum protective jacket to protect it for over a billion years in space, contains instructions on how it is to be played, with the cartridge and needle provided. At their playing speed of 16⅔ rpm, the records last for nearly two hours. On them are spoken messages in fifty-five languages, samples of Earth sounds, and a selection of music.

The recordings begin with 115 pictures which are encoded electronically into the grooves. Included among the pictures are views of our solar system, diagrams of human anatomy, landscapes of Earth, pictures of various human activities, and examples of human technology such as aircraft, radio telescopes, and space vehicles. Also electronically encoded is a written message from President Carter. Then follow the spoken greetings from around the world and what is described as a "sound essay" on the evolution of the planet Earth, from the sounds of volcanoes, rain, and surf (all important factors in the formation of life) to animal noises, the emergence of humans (denoted by footsteps and heartbeats), and ending with sounds of technology from the Morse code to the launching of a Saturn V and the beat of a pulsar. Most of each record is taken up with ninety minutes of music in twenty-seven excerpts ranging from pan pipes, a Navajo Indian chant, and an Australian horn and totem song, through Bach, Beethoven, and Stravinsky, to Louis Armstrong and Chuck Berry. The result should intrigue, or baffle, civilizations throughout the Galaxy in millions of years to come.

The Voyagers are not scheduled to leave the solar system until 1990, traveling at 17 km/sec; ground controllers anticipate that they should still be able to communicate with the probes, if they remain operational, until about 2007, when they will be 15 billion kilometers distant.

One day we might intercept a similar probe borne culture capsule from another civilization and hear what another planet sounds like, or gaze upon pictures of its inhabitants going about their everyday lives. The Pioneer plaques and the Voyager records, like the Arecibo message, are best seen as examples of what might be coming our way, rather than serious attempts at making our presence known to extraterrestrials, for the chances of their being picked up are almost unimaginably slight.

If the Cyclops radio telescope array described in the previous chapter is ever built, we would have a marvelous instrument for signaling to the stars, as well as listening. And in fact, it has been suggested that the forest of 1,000 or more giant antennae making up Cyclops be used to beam out messages in between listening projects, each individual antenna being aimed at a separate target. A 1-Mw beacon beaming through one of the 100-meter-diameter Cyclops dishes with 1-Hz bandwidth could be detected by the 76-meter radio telescope at Jodrell Bank (Cheshire, England) over 150 light-years away—*if* it were listening in the right direction at the right time on the right frequency. At 0.1-Hz bandwidth the range increases to 500 light-years, whereas another Cyclops could detect the 1-Hz beacon at an astounding 8,000 light-years.

If the entire 1,000-dish Cyclops were used en masse to beam a beacon at one star, assuming a total power of 1,000 Mw at 1 Hz, its signal could be detected by a suitably tuned omnidirectional antenna at over 150 light-years. Thus Cyclops could be picked up at considerable distances by antennae that were not even aimed at it. The Soviet SETI researchers are using just such omnidirectional antennae to listen for incoming beacon signals, and so they stand a chance of success if there are any Cyclops-like transmitters pointed our way.

Britain's Astronomer Royal, the radio astronomer Professor Sir Martin Ryle, in 1976 joined those who warn against the possible dangers of signaling to the stars. Yet there is no way in which we can keep our existence secret from those who are determined to overhear us. Since the 1940s powerful microwave beams from radar and television transmitters have been leaking out into space. This leakage noise is already slowly washing over the nearest stars like a tide of electromagnetic flotsam. Anyone listening to the right wavelengths with a sufficiently sensitive receiver within a range of about 40 light-years will already know that we are here, and would by now presumably be flagging for our attention. In the past decade or more, usage of the radio spectrum has moved to ever higher frequencies (UHF TV wavelengths range from about 70 cm to 35 cm) and to greater powers, so we have certainly given ourselves away, and will continue to do so for at least the foreseeable future. No one can be sure how long such profligate transmissions will continue, but probably they will eventually be replaced by the more economical direct transmissions from satellites, with consequent reductions in leakage.

TV and radio masts do not transmit omnidirectionally. Rather, they are constructed so that their beam is concentrated horizontally, where it is needed; therefore, it goes much farther for a given transmitter power. The signals received by extraterrestrials from domestic leakage will be fragmentary, since the rotation of the Earth will regularly sweep the transmitter beam across the extraterrestrial receiving antenna for perhaps only twenty minutes twice a day. But a typical 50-kw TV transmitter could be detected by a Cyclops system or its equivalent, such as a large space antenna, at a distance of about 50 light-years; since the same TV channels are shared by several transmitters around the world, the total power being leaked to space increases and the range of likely detection is correspondingly extended to perhaps 100 light-years.

Although it is unlikely that anyone will be able to decode our TV programs, the aliens will at least get an insight into how we split up the radio spectrum for domestic broadcasting, and they will also be able to judge our level of military activity from the extent of our radar surveillance. Perhaps the most powerful regular transmitters on Earth are those of the ballistic missile early-warning radars employed by the military of both West and East. Scottish science writer Duncan Lunan has suggested that some form of message to extraterrestrials could be superimposed on the radar beams. Wider radar coverage of the sky is given by the Space Detection and Tracking system of NORAD (North American Air Defense Command), which keeps track of all objects orbiting Earth.

Even without a deliberate message embodied on it, radio and TV leakage will reveal the approximate technological level of our civilization. The discovery of domestic transmissions also solves the wavelength problem for the other civilization: they simply beam in to us at the wavelengths already being used. In 1977 hoaxers broke into a news program from Southern Television in England with a "space message" purporting to come from an interstellar craft circling Earth. One day such a message may be received that is not a hoax.

In some ways, we have more chance of being detected by our leakage radiation than from the occasional transmission bursts suggested for the Cyclops array, because our domestic broadcasts are continuous. Professor David Bates of Queen's University, Belfast, a leading critic of SETI, has shown quite convincingly that flashing a message in turn at each of a large number of stars in the hope that someone is listening in our direction leads to impossibly long times before contact is established—perhaps 10 million years or more. The problem is to get the transmitting and receiving antennae pointing in the same direction at the same time.

Radio astronomers Gordon W. Pace and James C. G. Walker have introduced the idea of time markers for interstellar communication—pre-set instants when transmitter and receiver can simultaneously be turned toward each other. Since there can be no prior consultation between the transmitting and receiving civilizations, a time marker needs to be some event clearly observable to both parties, and Pace and Walker note that just such events occur in binary star orbits when the two stars are at their nearest to each other or farthest apart. The radio astronomers therefore suggest that civilizations orbiting a star in a binary system would choose to transmit at the moments when their parent stars were nearest together or farthest apart, and that we should accordingly listen or transmit to binary stars at those observed times.

As another solution to the problem of when and where to send signals, an American astronomer, William I. McLaughlin, and a Soviet scientist, P. V. Makovetskii, have independently suggested using a starburst, like a bright nova, as a time marker for interstellar transmissions. As an example, they cite Nova Cygni, the brightest stellar outburst for 33 years, which we on Earth saw in 1975. If other civilizations began to transmit signals as soon as they saw the nova, we can predict when their messages should arrive at the Earth. On this assumption, signals from Barnard's star would have begun to arrive on September 15, 1978. Signals from a star very similar to the Sun, known variously by its catalogue numbers of Gliese 788, DM +66 1281, or GC 28252, and lying 48.7 light years away in the constellation Draco, would have been expected to arrive beginning August 1978. Signals from Lalande 21185 would be expected in March 1984, from Tau Ceti in January 1987, and from Epsilon Eridani in June 1988. NASA's SETI project could be beginning at a very propitious time.

To ensure that contact is made in the shortest possible time, it seems clear that full-time transmissions to all stars of interest coupled with a large-scale listening project is necessary. One suggested mode of transmission is to have 1,000 or so individual aerials, perhaps like the Cyclops array, trained full time on each of the 1,000 most likely target stars. But this seems to me an inefficient and wasteful system. A large and expensive array (or rather, two, for we would need one on each side of the Earth to keep track of each star for twenty-four hours a day) would be devoted entirely to a very limited task, with no guarantee that the right stars had actually been selected. I cannot see us, or any other civilization, using such a system for a long-duration signaling project.

The only way to be sure of covering every possible target is to signal omnidirectionally. It is usually assumed in the SETI literature that some highly advanced civilization is transmitting with an omnidirectional antenna of enormous power, 1,000 Mw or more. But as we have seen, it would take a complete Cyclops system to detect such a beacon at distances out to 500 light-years, even with the quite narrow bandwidth of 0.1 Hz. What happens if civilizations are farther apart than this? With a 100-meter receiver, the range is only a pathetic 25 light-years. Therefore, with current radio astronomy equipment, a

1,000-Mw omnidirectional beacon is undetectable at the likely distances of other civilizations.

Since the intensity of a radio beam falls off with the square root of the distance, to increase the range ten times means increasing the power 100 times, in this case to 100,000 Mw. Again, I do not foresee that anyone would be using such an immensely powerful omnidirectional beacon, for quite practical reasons: no antenna would be able to withstand the immense voltages involved. It would rapidly break down and put itself out of action.

My argument is that omnidirectional beacons of practicable power are probably not detectable over the kind of distances that separate civilizations; they are certainly not detectable with our level of technology. Therefore, if we do detect an interstellar signal, or if we construct a practicable beacon ourselves, it will employ a different strategy altogether—one so far not seriously considered by signal searchers. I believe there is a more efficient transmitting strategy than either the fixed-dish or omnidirectional system that, in a limited form, we could begin today. In fact, I suggest that it may actually be easier to transmit signals to most stars in the Galaxy than it is to receive them, thus placing the onus for advanced technology on the receiving civilization rather than the transmitting one, which is an interesting inversion of the usual assumption.

Should the current searches for messages from the stars described in Chapter 7 draw a blank, I believe that we should consider setting up a long-range transmission system ourselves, on the assumption that someone is waiting for a lead from us. Interstellar communication can never succeed if everyone listens and no one transmits! My own feeling is that the lack of signs of any other life in the Galaxy to date indicates that other civilizations, if they exist, are sparsely spread and that we may need to make ourselves heard over thousands of light-years. Perhaps we will never be accepted into galactic culture until we choose to announce our existence. Our first message to the stars will be like the first cry of a baby, announcing our birth.

Nature has simplified the transmitting problem for us, because most of the stars in which we are interested lie in the narrow band of the Milky Way, which is centered on the plane of the Galaxy. Therefore we could transmit to most of the stars in the Galaxy by sweeping a

number of dishes needed at the start, we might spread the beam out wider in the hope that there is a Cyclops-type collector available to detect us.

By trading off a small amount of time, and not being greedy in the amount of sky that we attempt to cover, we can therefore signal to the stars with almost as much chance of being detected as with continuous all-sky coverage and without recourse to implausibly high-powered omnidirectional beacons.

The drawbacks with such a ground-based system are the mechanical problems involved in continually driving radio dishes along the galactic plane, and also the power needed—although this latter problem is shared by all beacon-transmitting systems. Both these problems are solved by transferring the lighthouse-beam system into space.

The idea of a spinning signal beacon in space makes particular sense because once an object in space has been set spinning, it continues unhindered. In space, the entire sky is visible from one spot, which eliminates the need for duplicating the system as is necessary in each hemisphere on Earth; and, of course, there is plentiful free power from the Sun. Construction of an interstellar beacon seems an ideal task for the O'Neill colonies.

The beacon would consist of a row of antennae studded along the outside of a spinning cylinder. Since the cylinder is not intended to be inhabited, except perhaps for a residential scientific team or occasional maintenance crews, its diameter can be as small as practicable—say 100 meters wide. A cylinder of such small diameter spinning once every few minutes generates only a fraction of Earth gravity, so the antennae can be lighter and simpler than in a comparable Earth-based system. The cylinder's axis will be aligned on the galactic poles, so that the beam from the transmitting dishes sweeps the Milky Way on each rotation. If we assume, as before, 200 dishes of 100-meter aperture each, the total length of the cylinder required is 20 kilometers, less than the Island Three design of O'Neill.

The power station for the interstellar-signaling cylinder will be similar to that used in the solar power satellites supplying energy to Earth. The 10,000-Mw total output of such a station, divided among 200 antennae, gives 50 Mw per dish, sufficient for the transmissions to be picked up by a Cyclops at the galactic center. A beacon colony would be best sited as far away from Earth as possible, both to reduce interference on Earth and to prevent Earth from blocking the beam from part of the sky. Probably the best place would be on the opposite side of Earth's orbit, behind the Sun.

A fully rotating 360° lighthouse beam in space would produce flashes like a slow pulsar. By the next century we could be operating such a beacon, and it does not seem implausible to imagine that civilizations only slightly more advanced than ourselves are already employing such a system. Civilizations of a technical level equivalent to our own could already be using the Earth-based version. I therefore conclude that interstellar signaling is possible for civilizations at our level of technical development, which considerably increases the number of potential messages from the stars.

An understanding of the likely strategy of a transmitting civilization gives us a better idea of what to listen for in our own searches. Although I believe that many civilizations like our own lying in the galactic plane could be using such a system at this moment, their outputs would be more difficult to detect than is normally anticipated for interstellar transmissions, which helps reconcile the fact that no such transmissions have yet been discovered, either by deliberate attempt or accidentally.

However, we can detect these signals once we know what to listen for. We should therefore organize an international radio sky-survey project, like the Carte du Ciel of optical astronomy in the last century, with each nation allotted an area of sky whose solar-type stars they should survey at likely wavelengths for 1,000 seconds or so (15 to 20 minutes) at a time. They should be on the lookout for spikes of energy every few minutes as the lighthouse beam illuminates us briefly. After rechecking on each star for six months to a year, to take account of possible eclipse of the transmitting planet by its parent star, we would know whether there were any beacon signals to be picked up. The searchers should be encouraged in their task by the fact that the requirements of interstellar signaling are clearly not as extreme as some pessimists have declared.

There may be many weak radio pulsars with periods of a few minutes that are actually lighthouse interstellar communication beacons being transmitted from O'Neill-type colonies around other stars, but they would have gone unnoticed to date because radio observatories do not scan for pulsar periods longer than a few seconds. If such transmissions were detected, their artificiality would be immediately evident, because natural pulsars do not transmit that slowly, and also because of the very narrow bandwidth of the beam.

All SETI projects are based on the faith that communicative civilizations exist simultaneously in the Galaxy. We must proceed on this faith, or not proceed at all. While our efforts are confined to simple listening projects there is no great commitment, but once a civilization begins a signaling project it is committing itself to an essentially open-ended task. Not only must we face the prospect of signaling endlessly, but we should also listen endlessly too, in case our efforts have triggered a return message. At least the listening can be sporadic, since a reply is likely to be beamed full-time at us for a long period. Once the target has been singled out, all available power can be devoted to sending a complex message of the type described at the end of the last chapter.

An endless SETI commitment may sound daunting at first, but it is no different from many of the other commitments inherent in our society. For instance, public utilities such as electric companies face the fact that their task will continue endlessly, generation after generation, century after century. When we invest with a bank we assume (I hope rightly!) that we are dealing with a permanent institution. Many such businesses are commitments to the permanence of civilization.

A SETI signaling project is a statement of faith not only in the permanence of our own civilization, but also in the permanence of others.

REFERENCES

Bates, D. R., *Nature*, Vol. 248 (1974), p. 317.
——— *Nature*, Vol. 252 (1974), p. 432.
——— *Proceedings of the Royal Irish Academy*, Vol. 77A (1977), p. 45.
Lunan, D. A., *New Scientist*, Vol. 74 (1977), p. 490.
McLaughlin, W. I., *Icarus*, Vol. 32 (1977), p. 464.
Makovetskii, P. V., *Soviet Astronomy*, Vol. 21 (1977), p. 251.

National Astronomy and Ionosphere Center Staff, *Icarus*, Vol. 26 (1975), p. 462.

Pace, G. W., and J. C. G. Walker, *Nature*, Vol. 254 (1975), p. 400.

Ridpath, Ian, *Nature*, Vol. 253 (1975), p. 230.

——— *Journal of the British Interplanetary Society*, Vol. 31 (1978), p. 108.

Sullivan, W. I., S. Brown, and C. Wetherill, Science, Vol. 199 (1978), p. 377.

9

Messenger Probes

Somewhere in space, a message from the stars may be on its way toward us in the electronic brain of a messenger probe sent by another civilization. This is the visualization of Stanford University radio astronomer Ronald Bracewell, who believes that space probes are better under certain circumstances for initiating interstellar communication than a radio beacon.

His argument runs as follows. If life is plentiful—hundreds of millions of civilizations throughout the Galaxy (call this Case I)—then there is no problem in establishing first contact by radio, because civilizations would be no more than 30 light-years apart and would each need to scan only about fifty stars before they found evidence of another's existence. Unless we have been particularly unlucky so far in our listening projects, there seem to be no civilizations this close to the Sun, and we can conclude that the Universe is not inhabited to such a high density.

Therefore we must be dealing with civilizations that are, at best, sparsely distributed in the Galaxy: say, from 30 to 300 light-years apart. This, in Bracewell's analysis, corresponds to Case II, where there are perhaps 10 million civilizations in the Galaxy. In Case III, where life is rare (only a few thousand civilizations), civilizations are so far apart they hardly ever live long enough to exchange a message—

although some distant race might occasionally emit a swan song to which it expects no reply.

If Bracewell's Case II corresponds to reality, then a direct radio message is not necessarily the best way of setting up communication. Beyond the 30-light-year range the number of potential target stars increases dauntingly, to 1,000 within 100 light-years' range, and 10,000 within 300 light-years. The implications of these figures apply to all forms of signaling and reception, and are worth considering in detail.

For two civilizations to make contact, one must be listening at the same time as the other is transmitting. This part of the problem can be solved if the transmitting civilization aims to cover all likely target stars with either a permanent beam or a rapidly repeating signal that will be picked up with only a few minutes' listening, as discussed in the previous chapters. Oddly enough, our reception of a beacon signal would put us at an advantage over the senders: we know where they are, but they do not know where we are. Should we decide to reply, our message of acknowledgment will be arriving at a time unknown to them, from a direction that they are equally unable to predict. Therefore, they must put as much effort into listening for a reply as they do into sending the initial signal—an effort that could last centuries at the very least.

Even assuming that the transmitting civilization is prepared to undertake such an open-ended project, there would be political problems at our end. If the civilization is, say, 100 light-years away, two centuries will elapse before we know whether our reply has been picked up: our message, the equivalent of "Hello, how are you?" will take 100 years to cross the gap, while the other civilization's rejoinder, perhaps something like "Good to hear you, I'm fine," will take another 100 years to arrive. In the meantime, we might as well transmit something along the lines of "Let me tell you more about myself" while we are waiting. But a major project of this sort that will tie up a large radio telescope and considerable amounts of power for centuries on end may not commend itself to legislators on Earth. Clearly, the whole process of establishing radio communication over distances of hundreds of light-years is somewhat cumbersome, and is open to pitfalls on both sides.

Bracewell believes that these problems can be avoided by the use of intelligent automated probes as intermediaries in establishing first contact. In principle, it is possible to pack an enormous amount of information into an interstellar probe and send it out toward a likely star. When such a robot messenger encountered another civilization it would be able to converse with them while it relayed news of its discovery across the 100 light-years or so to its parent planet. The probe would tell the receiving civilization where to point its radio telescopes so that both civilizations could enter into direct contact with the minimum of wasted effort.

The sending civilization, which Bracewell terms Superon, would naturally need to be patient while its probes reached their targets and awaited the emergence of communicative life. And there would need to be as many probes as there are target stars—up to 10,000 of them within a 300-light-year limit. In the long run, though, it might be more economical to follow this plan than to squander endless power on an unheard beacon and waste uncountable hours on searches for a nonexistent answer.

Possibly, Bracewell hazards, our nearest galactic neighbor is already sending out messenger probes, with the Sun as one of its targets. However, if no contact has been established by the time we begin to explore the stars, it will be our turn to assume the more active role.

Let us follow the fortunes of one probe, sent to investigate a star 100 light-years or so from Superon. We assume that its propulsion unit is similar to that of the Daedalus starship described in Chapter 6, so that it travels at one-tenth the speed of light and has a journey time of 1,000 years. When the probe arrives at the star, it will swing into orbit within the star's habitable zone. Once there, its instruments may be able to distinguish the most likely planet to support life. Drawing its power from the light of the alien sun, the probe will settle down to listen for the awakening of technological civilization on that planet. The simplest way of detecting technological life on a planet is to listen for leakage radiation from its domestic broadcasts. These would be too faint to be detected by radio telescopes over distances of several hundred light-years, but an on-the-spot probe would pick up the noise easily. Of course, even if the planetary system proves sterile the probe will not have been wasted, as its instruments can make scientific observations like those planned for Daedalus.

On arrival the probe sends back its preliminary report on the alien solar system to Superon. Then follows silence from the probe, possibly broken every century or so with a progress report and telemetry on the state of its on-board systems.

But one day the probe's ever-listening radio ear detects a stirring in the electromagnetic spectrum on the planet within the life zone. This is the sign it has been waiting for. The probe begins to activate its dormant circuits, updates its view of its surroundings, and attentively studies the first chirps of this fledgling civilization.

The first sounds come only sporadically, and at the long-wavelength end of the radio spectrum. But the probe can afford to be patient a few more years yet. Soon, regular broadcasts can be heard weakly filtering their way into space. And then comes a major advance: the transmissions, now much stronger, move into the high-frequency region of the spectrum. Sudden bursts of high power and complex signals carried on very high frequency bands indicate that the civilization is developing radar and TV. It is time for the probe to move into action. How should it respond?

Here, the frequency-selection problem is simple. The probe simply beams out a signal on one or more of the frequencies that it knows are already in use on the planet below. Somewhere there must be a receiver already tuned to those transmissions. Possibly, the probe's transmissions would be received on domestic FM radio and TV sets.

What would be the nature of the probe's transmissions? It could beam pure interference, a kind of jamming signal that is bound to be noticed—but which would cause inconvenience, consternation, and not a little panic as radio engineers quickly traced the signal to a source in space not far from their home planet. Ronald Bracewell, though, suggests another strategy. Rather than interrupt the transmissions, why not amplify and replay them, thus giving an echo effect? Receivers on the planet would pick up first the direct transmissions, and then the signal retransmitted from the probe, the delay time being seconds or minutes depending on the probe's distance. This strategy would not only appear more friendly than an indiscriminate jamming, but it would also indicate that the probe was prepared to interact with the transmissions, thus providing a basis for two-way communication.

Once the probe's existence is clear and its position in space has been located, radio engineers on the planet below can begin to converse with it. But first they must change the character of the transmissions which the probe is echoing, so that the probe knows it has been spotted. They would probably send short radio bursts toward the probe, interspersed with pauses in which it can make its reply. After a short to-and-fro exchange of transmissions and echoes, it should be clear to the probe that it has caught somebody's attention.

Next the probe will want to know more about the civilization's technical capabilities. By progressively weakening the strength of its echo and noting when the civilization ceases to respond, the probe can deduce the sensitivity of the equipment it is dealing with. It can also slowly begin to change the frequency of its transmissions and see how long the civilization can follow it. By slowly changing the frequency of their own responses, the civilization can guide the probe onto the wavelength that is most convenient for them, allowing the domestic radio channels to return to their normal business. By then, of course, the probe will already have gained a fair insight into the technological level of the civilization and will have begun to flash news of its discovery back to the Superon Space Agency, which will be alert for such stop-press news from any of its messenger probes.

Now comes the time for the probe to unfold the messages that it has carried in its computer brain across the 100-light-year gap from Superon. Even without knowing the recipients' language, the messenger probe can put across plentiful information about its home and its creators in the form of pictures. The first interstellar message will therefore come in the form of television, a universal sign language. The probe will either transmit a test pattern until it receives an acknowledgment that the signal is being successfully reconstructed below, or it might obligingly encode its transmissions so that they fit in with the existing TV systems.

Bracewell speculates that the first picture from the probe will be of a constellation of stars familiar to the receiving civilization, followed by a zooming in on the home star. This is the crucial part of the message: it directs attention to the source of future radio transmissions. Another important part of the probe's message will be the correct wavelength and strength of the interstellar signal to be sent by Superon. Of course, no transmissions will be sent until Superon knows that the probe has found life, but the recipient civilization will know how long they have to wait by measuring the distance to the probe's home star. In the meantime, the civilization can develop the technology they will need, or even flash messages of their own.

As the probe's fantastic TV travelogue continues, the home star will become visible as a disk, perhaps with starspots. Around it a family of planets will swing into view until at last the picture zooms in on Superon, the home planet itself. What follows is restricted only by the imagination of the probe's senders and the ingenuity of the recipients in reconstructing it. Perhaps the message might include digitized holograms, reconstructable like pictures received from our own space probes, which will give a three-dimensional picture of the Superonians, their dwellings, examples of their technology, and their works of art. Alternatively, actual holograms may be carried aboard the probe; part of its message may be an invitation to retrieve it from orbit to obtain artifacts stored on board.

The probe's memory banks may contain a simple pictorial dictionary to help the recipients learn Superese, the language of its makers, and to aid it in translating the language of the planet below—and that planet might be Earth. Says Bracewell: "Knowledge of our language will enable the probe to tell us many fascinating things: the physics and chemistry of the next 100 years, wonders of astrophysics yet unknown to man, beautiful mathematics. After a while it may supply us with astounding breakthroughs in biology and medicine. But first we will have to tell it a lot about our biological make-up. Perhaps it will write poetry or discuss philosophy. Perhaps the messenger knows how the Universe started, whether it will end, and what will happen then. Maybe the probe knows what it all means, but I wonder . . . I think that is why Superon wants to consult us!"

The probe's ability to react quickly to the situation below will help circumvent any political disputes that its presence might spark off. For instance, some governments might wish to keep the probe's information to themselves, and jam the signal if they thought rival governments were likely to gain its secrets. A dumb radio message arriving from afar would be open to such treatment. But the probe knows better. First, it can select who it talks to by dropping its power to such a level that only an international organization such as the deep-space tracking network of NASA could keep in contact with it. Secondly, it can move frequencies to prevent jamming. And it can, if it wants, initiate communications with any transmitting station on the planet. The circumvention of rivalries between nations would be uppermost in the probe's pre-programmed strategy.

The messenger-probe concept clearly overcomes the problem of putting two civilizations in contact. A civilization contacted by a probe can be fairly sure that any response it makes to the home star will be picked up; and it will know that, once the probe's positive report has reached home, there will be more transmissions on the way. What's more, Superon is unlikely to be the only advanced civilization within our range; the probe may tip us off about a number of other communications centers with which we may swap messages. "There will likely be a galactic club, whose members are experienced at finding developing communities such as ours and inducting them into the galactic community," declares Bracewell.

Unlike a radio-beacon signal, the probe strategy can be tailored to fit the prevailing economic climate on Superon. Only one probe a year or even less need be sent; the entire program can be stopped temporarily without upsetting the eventual chance of success, so long as there is always a monitoring station in operation to pick up reports from the probes as they reach their destination. The one drawback is the long time before Superon completes its project; the first report will not come back for centuries, and the final probe will not be launched for millennia. But of course a radio-beacon project must also be measured in centuries or millennia. This is a fact of life; an interstellar culture must inevitably think in far longer time scales than we are presently used to on Earth. Bracewell's point is that the probe strategy should in the long run be more successful than sending radio beacons *if* life is reasonably frequent throughout the Galaxy.

Is it possible that such a probe entered our solar system long ago and has been orbiting silently ever since between Venus and Mars? The existence of such a probe, should we ever find it, would be mute but incontrovertible evidence of other beings in space. If there are artifacts on board, designed to survive long after the probe's electronic systems have decayed, we may still learn much about Superon and its inhabitants. If the Superonians were merely keen on spreading life around the Galaxy, they would have crashed the probe on the most habitable-seeming planet (presumably Earth) and released a wide variety of biological samples in the hope that some would take root. There is no evidence of any alien life forms on Earth.

But could there be a probe out there today, not dead but very much alive and trying to contact us?

One interesting case concerns TV station KLEE in Houston, the test pattern of which was allegedly received on TV sets in England in 1953, years before the advent of communications satellites or even the first Sputnik. What made the achievement even more remarkable was the fact that KLEE-TV had changed hands three years previously and was by 1953 transmitting under a different call sign, KPRC.

This fascinating case has been thoroughly investigated by radio astronomer Frank Drake, and the explanation is quite clear: it was a hoax which backfired. The hoaxers were two English businessmen who were demonstrating TV sets which they claimed could pick up overseas TV. Strangely, all that ever showed up were call signs—no programs or moving pictures at all. The hoaxers encouraged gullible witnesses to photograph the signs, including one supposedly from Moscow written in English, and send the photographs to the TV stations for confirmation. Many examples of these photographs are in the possession of station KPRC in Houston, along with considerable correspondence on the matter. The photographs reveal the call signs to be poor forgeries; presumably the signs were projected inside the actual sets by the hoaxers. When the hoaxers made the mistake of projecting an out-of-date call sign, their deception should have been immediately exposed. Instead, it provided another UFO-type mystery for the gullible.

A far better example of the possible "echo" effect of a messenger probe, and one quoted by Bracewell himself, was the series of curious radio pulses picked up in the late 1920s during early experiments on long-distance radio propagation. In particular, several remarkable sequences of echoes were recorded in October and November 1928 from transmissions made by station PCJJ in Eindhoven, the Netherlands. Echoes of varying delay, from three to thirty seconds, were picked up in Oslo, London, and Eindhoven itself. On the face of it, there seemed no explanation for echoes of such long delay; even a reflection from the Moon would have given a delay of only 2.5 seconds, and the echoes would have been far weaker than those actually recorded. One of the remarkable facts about the long-delayed echoes, or LDEs as they came to be called, was that they did not get fainter with delay time, as would be expected with a simple reflection effect. So whatever was delaying them was also amplifying them.

Since then, radio operators have continued to pick up LDEs sporadically, and the phenomenon has become known as radio's flying-saucer effect. Explanations centered around plasma effects in the ionosphere.

Then in 1972 a Scottish writer, Duncan Lunan, decided to take Bracewell's space-probe suggestion more seriously. When he plotted a series of LDE pulses from Eindhoven transmissions recorded by Jorgen Hals and Carl Stormer in Oslo on October 11, 1928, he found what appeared to be a star map of the constellation Boötes, the Herdsman, with particular attention drawn to the star Epsilon Boötis. Lunan tentatively concluded that the LDEs were being sent by a probe which had come from Epsilon Boötis and which now lay at one of the gravitationally stable Lagrangian points of the Moon's orbit; analysis of LDE events showed that they occurred most often when the Lagrangian points were high in the observer's sky.

This interpretation created considerable interest when published in the magazine *Spaceflight* in 1973. The following year Lunan published interpretations of further LDE "dot pictures" dating from 1928 and 1929 in his book *Man and the Stars*, supporting the idea that the probe was sent by inhabitants of a planet around Epsilon Boötis.

Epsilon Boötis is a double star, the brighter component of which is an orange giant, swelling up at the end of its life. In Lunan's scenario, the probe was one of many that had been sent out as a last dying gasp by the inhabitants of Epsilon Boötis in search of a new home. However, Lunan later abandoned this interpretation when it became clear that Epsilon Boötis was approximately twice as far as previously believed, which meant that the component stars were both bigger and brighter than previously assumed, and thus too short-lived for life to have arisen in their vicinity. Abandoning Epsilon Boötis as the theoretical star of origin also invalidated the other "dot picture" interpretations. Lunan has not tried to reinterpret these.

Instead, he has since claimed that Epsilon Boötis would be a prime navigational reference for a starflight from Tau Ceti to the Sun; as seen from Tau Ceti, our Sun would lie in Boötes. In 1975 a Russian astronomer, A. V. Shpilevski, published an alternative interpretation of the October 11, 1928, LDEs in the Polish magazine *Urania*. By plotting the same dot series in a different way he found a star map of the constellation Cetus, with the star Tau Ceti indicated as the star of origin. Thus, if the star map has any reality, it shows us that the probe came from the nearby star Tau Ceti, from which no radio noise has ever been detected. However, as we shall see in Chapter 15 apropos another disputed star map, when the same data can be used to "prove" two mutually exclusive interpretations, this indicates that both results are illusory.

Further problems for the space-probe theory came when Tony Lawton, an English electronics engineer, traced records of LDEs from the Eindhoven transmissions made by Edward Appleton and his colleague R. L. A. Borrow at Kings College, London, in November 1928 and February 1929. The Appleton-Borrow LDEs do not yield a star map. It became clear from these records that not every transmitted pulse produced an LDE, whereas some pulses produced a train of LDEs; if the LDEs received on October 11, 1928, in Norway were not a continuous sequence but had followed a similarly fragmentary pattern, this would be sufficient to destroy the star-map plot. What's more, the LDEs Appleton and Borrow received were a mixture of clear and blurred echoes, which does not sound like the deliberate transmissions of a probe, but could well be the result of natural effects.

Lawton conducted a search for LDEs during 1974 and 1975 with a colleague, Sidney Newton. Most of their work was confined to listening for echoes of existing transmissions, but they also undertook six sending attempts of their own, three using Morse signals and three by transmitting a two-tone frequency that would have been unmistakable had it been echoed. They found that LDEs are very rare; they did not receive one certain LDE during their entire vigil. If long LDE sequences exist like those of 1928–29, they are swamped by other transmissions that flood the airwaves today.

Lawton and Newton developed a natural explanation for LDEs. They propose that LDEs are caused by plasma effects not only in the ionosphere but also in regions around the Lagrangian points, which can occasionally fill with dust and ionized gas; visual observations of faint glows around the Lagrange-point areas have been reported. With such a mechanism, the radio waves would pass right out of the ionosphere and into the gas around the Lagrange-point areas (the Trojan ionosphere, as they term it) before being reflected. Effects in the plasma would account both for the delay times and the amplification of the signals. This would also explain the association of LDEs with the position of the Lagrangian points in the observer's sky.

Although Lunan himself now doubts the probe's existence, he has not altogether given up hope. He maintains that the only conclusive test is to retransmit the star-map pattern to the Lagrangian points of the Moon's orbit, which ought to reactivate the probe if it exists. Accordingly, with colleagues in ASTRA (the Association in Scotland to Research into Astronautics), he plans just such a decisive transmission. "If only," as he says, "to bring the speculation to a close." But even by the end of 1978, these transmissions had not taken place.

REFERENCES

Bracewell, R. N., *The Galactic Club* (San Francisco: W. H. Freeman, 1975).
Lawton, A. T., and S. J. Newton, *Journal of the British Interplanetary Society*, Vol. 27 (1974), p. 907.
Lunan, D. A., *SERT Journal*, Vol. 10 (1976), p. 180.

10

Consequences of Contact

Humanity may be united by the reception of that first message from the stars, or it may be irrevocably split. Or, more likely, it will carry on much as before. Over the centuries, civilization has developed a remarkable resilience to new ideas. Copernicus in 1543 produced the major bombshell of the Renaissance when he asserted that the Earth was not the center of the Universe but was instead an ordinary planet orbiting the Sun; this was confirmed by the calculations of Johannes Kepler and the observations of Galileo Galilei. Intellectuals and churchmen were thrown into a turmoil, but normal everyday life, as far as can be determined, continued unaffected. In the last century, the evolutionary ideas of Charles Darwin created a scandal in high society, but were soon assimilated into mainstream thought. In the 1920s, perhaps the most astounding astronomical discoveries of all—that ours was but one Galaxy in a Universe full of other galaxies, and that the Universe as a whole was expanding—passed almost without a ripple. Nowadays, people are actually bored with the idea of men walking on the Moon.

There has been so much cultural conditioning that if extraterrestrial life were eventually discovered it would come as an anticlimax rather than a shock. It is the unknown and the unexpected that cause the greatest surprise, and extraterrestrial contact is not unexpected. Extraterrestrial beings already exist, at least in many people's minds. Humans seem eager to grasp the concept of extraterrestrials, as though the possibility of mankind being alone has now become the more frightening alternative.

I recall joining an audience of several hundred people at Caxton Hall, London, in 1973 to hear Duncan Lunan describe what he believed might be a radio message from an alien space probe (see previous chapter). The scene reminded me of nothing so much as Professor Challenger's public address on his return from the Lost World, in Conan Doyle's book of the same name. The press had heard of Lunan's ideas, and the meeting had been widely publicized. But there was no panic; no hordes tried to force their way in, and the audience remained calm throughout. A similar religious orderliness prevails at UFO group meetings. From this I can only conclude that people are either thoroughly prepared to meet extraterrestrials, or that they don't believe what they are being told!

When we are faced with incontrovertible evidence of extraterrestrial contact, of course, the situation may be different.

We must distinguish between indirect contact, such as by radio, and direct physical contact, for we react to these differently. A long-distance radio message from the stars would seem relatively innocuous. We are accustomed to receiving graphic new cultural and intellectual inputs via TV and movies—views of far-off lands, strange people, and curious customs—but are very little affected by them. Personal contact is something different altogether. In 1938 the famous Orson Welles broadcast of *War of the Worlds* created panic because it seemed as though a physical contact with extraterrestrials was occurring, and a hostile one at that; people perceived an immediate threat, which is not so apparent in a message. In a way, we have already experienced one personal contact which has radically changed the Western world: the arrival of Jesus Christ. His philosophy had more impact *at the time* than the more slowly assimilated ideas from Copernicus and Galileo onward that have recast our views of the Universe (Jesus did not teach any astronomy, incidentally).

Whereas one would expect hordes of sensation seekers besieging the radio observatories at which searches for interstellar signals are being made, there is an almost total lack of curiosity. Perhaps people fear that scientific facts would spoil their fantasies.

What would astronomers do if they intercepted that first message from the stars? I have quizzed several signal searchers, and their replies reveal a surprising divergence of opinion. Benjamin Zuckerman and Patrick Palmer, who have used the Green Bank telescopes for the major survey to date, said that they did not initially give much thought to their actions should a message be discovered (an admission that they were not expecting to succeed?). Zuckerman says: "I think that we would try to confirm beyond a doubt that it was a CETI [communication] signal before we let the world know about it. I would also think very hard about replying to 'Them' before we made our results public."

Patrick Palmer says: "My desire has always been to handle it as a normal scientific discovery. Our greatest interest would be in understanding the message and communicating, and we would certainly need the aid of many other scientists to do this as efficiently as possible. I believe that it is essentially always wrong to try to withhold basic scientific information." He dismisses as "science fiction" any suggestions of government attempts to suppress positive results, noting that in three years of work on signal searching there had been no contact from any governmental agency. "Since we use a national facility [the National Radio Astronomy Observatory at Green Bank is open to qualified investigators from many sources] rather than our own observatory, keeping something secret is very difficult because of the sheer number of people passing through and the openness of the observatory operation."

Canadian signal searchers Paul Feldman and Alan Bridle say they feel the release of CETI information "depends very much on the apparent content of the message"—an indication of the worries that some scientists feel about the implications of interstellar contact. Carl Sagan, by contrast, says that his first response would be to ask for confirmation of the signal from other observatories, to establish its reality. "It would be very hard to hide the existence of a message."

Bernard Oliver has pointed out that dangers do not arise with the detection of a signal, only our response. Before we answer a CETI message or radiate a beacon of our own, he says, "the question of potential risks should be debated and resolved at national or international level." This echoes the guidelines laid down at the 1971 Soviet-American CETI conference at Byurakan, Armenia, where the participants declared: "It seems to us appropriate that the search for extraterrestrial intelligence should be made by representatives of the whole of mankind." By extension, this implies that any conscious signal transmitted should also be the subject of international agreement and cooperation, which was not the case with the Arecibo message (Chapter 8).

Free and orderly dissemination of information about an extraterrestrial contact will be vital to prevent undue confusion and alarm. A fine example of how to handle a potentially explosive situation was provided in 1957 by the Smithsonian Astrophysical Observatory after the surprise launching by the Soviet Union of the first Sputnik. Smithsonian scientists decided to tell the public all that was known; they set up an information center which held regular press conferences and at which experts were able to help newsmen sort out fact from fiction, truth from rumor. This open approach helped dispel much of the fear that the surprise launching inevitably aroused. The need to set up such a reliable information source is paramount should messages from the stars be intercepted.

In 1967 radio astronomers at Cambridge, England, were faced with a similar situation when they detected the flashing radio sources known as pulsars, which they at first thought might be interstellar beacons. But the astronomers did not make an announcement until they had studied the objects further and determined their true nature.* A botched announcement would have caused considerable misunderstanding, as happened in 1973 when Soviet astronomers revealed that they were picking up signals that might be from extraterrestrials (see Chapter 7). Western newsmen and scientists had difficulty getting more information following this preliminary announcement, which is now believed to have been due to a misinterpretation of natural signals and satellite transmissions. Such a false alarm and the subsequent apparent clampdown on information do not inspire confidence in the abilities of some nations to handle events should they be the first to detect a message from the stars.

There would doubtless be a wide range of reactions to the discovery of extraterrestrial life. Scientists will regard it as the greatest discovery ever made, outstripping in importance the release of atomic power and the landing of the first men on the Moon, two events whose implica-

* The Cambridge astronomers were still discussing what to do when it was proved that the signals had a natural origin.

tions are even now not fully realized. Discovery of lowly life forms, such as were thought to exist on Mars, would seem comparatively innocuous; but what about evidence of advanced technological life forms?

A radio message, for instance, will have its impact affected by the speed with which the signal is confirmed and decoded. A simple message could be understood immediately, so that the scientists could announce receipt of the first message from the stars and at the same time tell us what it said. But a more complex message could take months or years to decipher; we will first be told that an extraterrestrial message has been received, but only slowly will we learn of its contents as it is decoded and interpreted like an ancient script. This second chain of events would considerably cushion the blow.

Those with the most dogmatic beliefs will be the most affected by news of intelligent extraterrestrial signals. Flying-saucer cultists would claim it as final confirmation of their unsubstantiated myths; yet it will be nothing of the kind, for visitors do not telephone long-distance if they are already dropping in in person. If we were visited in person, it would allow us to compare real extraterrestrials and their genuine interstellar craft with the shimmering visions of the UFO evangelists. First contact with extraterrestrials may be to the UFO cults what Earth satellites were to the Flat Earthers.

But the profoundest response of all will be religious. People will take new stock of the world and their relation to it, as they had to after the realization in the seventeenth century that the Earth was not the center of the Universe. Whereas we know that the Earth has no privileged position in space, we do not as yet know that there are beings superior to us anywhere in the Universe; thus we tend to regard ourselves, albeit subconsciously, as the pinnacle of creation. Discovery of superior extraterrestrials would dent our ego, and we might react out of sheer annoyance at their existence. We would want to try to assert our imagined superiority over the extraterrestrials, particularly if they visited us in person.

Traditional religions will need to reexamine their assumptions. Western religions, in particular, with their focus on Jesus Christ, may be in the biggest trouble of all. Perhaps the major outcome will be a further demysticizing of religion, a move from the spiritual to the more material, a weakening of the concept of God in favor of an emphasis on humanism. A man of the cloth sitting next to me in a radio discussion on extraterrestrial life confessed that his main concern was with the quality of the alien life—that is, whether the beings were good or evil. But these are human concepts; such ideas might have no meaning to an extraterrestrial, and we must guard against judging the aliens by our own standards. This is particularly true in physical and social terms: a major shock could be something about their appearance or their behavior (a cannibalistic society, for instance) which is unacceptable or repugnant to us. We all have preconceived notions about the "right" behavior which are often shattered by contact with other cultures on Earth. How much greater will the contrast be with aliens? Cultures are shaped by their environment: thus other environments, other cultures.

It seems that interstellar signaling is the best way of initiating relations, because it removes the physical threat to both groups of beings. We are, for instance, suspicious of the uninvited caller, whereas one who phones first is more readily accepted. But we have no guarantee that this is how contact will be established. Instead, our first brush with extraterrestrials may be an unexpected personal visit, and this is where danger lies. The surprise encounter may come with the unintentional crash-landing of an alien ship or an accidental meeting with a survey team.

Certain dangers of a visit from extraterrestrials have been exaggerated, such as the possibility that they would carry us off for food. Anyone embarking on an interstellar mission would be sure to grow their own appetizing food on board, as in the O'Neill colonies, rather than rely on the off chance of finding something palatable at their destination. In fact, in view of the probable differences in biochemistry on different planets, one good rule would be not to eat anything on an alien planet. To avoid misunderstanding with the natives, another good rule would be not to go prospecting for natural resources in inhabited solar systems. But in any case I cannot see the value of prospecting the Moon, say, for shipment to Alpha Centauri or beyond; advanced civilizations will get their raw materials much nearer home.

One real danger, albeit unintentional, will be that of disease. On Earth, isolated communities have been decimated by relatively innocuous diseases such as influenza introduced by visitors; how much greater might the health danger be from visitors from other planets.

For their own safety as well as ours, alien visitors are unlikely to walk around on Earth without spacesuits, or allow us into their craft without similar precautions to prevent the exchange of germs.

One way in which visitors could meet us on more equal terms would be by making contact with a Moon base or space colony. That way, they could be sure of meeting a select, science-oriented group with similar experience of the space environment. With so much in common, rational communication would be eased, and conditions for an in-person meeting would be better than on Earth, for representatives of both groups could meet in spacesuits in the neutral territory of space or the lunar surface. Off-Earth contact would help buffer the shock of that first alien encounter, as long as it was carefully handled. Members of a Moon post might feel more apprehensive of the aliens, because of their isolation. Back on Earth, the reaction could be greater because of uncertainty about what was happening, which might lead to a well-intentioned but clumsy "rescue" expedition that could create unwanted confrontation.

We must assume that anyone who makes such an approach must be friendly, since ill-intentioned visitors would swamp us without introducing themselves, and there is nothing we could do about it. It would need just one spacecraft to poison the atmosphere and the seas to kill or immobilize all life on Earth. Invasion from space is one of those theoretical hazards of life, like possible extinction of living organisms by the supernova explosion of a nearby star.

Some people have urged that we should not transmit radio messages to the stars for fear that this would give our existence away to possibly aggressive extraterrestrials. But an interstellar King Herod who was intent on stamping out the newborn of the Galaxy would find us soon enough anyway. Since apes in a recognizably protohuman state were around at least three million years ago, it is reassuring that we have been allowed to progress this far without interference.

I therefore assume that any contact in which we are allowed freedom of action will be a peaceful contact, and it will be up to us to ensure that the contact continues peacefully. The difficulties are immense; imagine, for instance, the alarm of an alien crew surrounded by eager reporters with popping flashbulbs.

Anthropologists will be the most important intermediaries in a peaceful contact. They will need to explain our culture to the visitors, and discover more about the behavior and ethics of the aliens so that neither group offends the other by some unintended rudeness or sacrilege. If nothing else, it will be illuminating to see ourselves as others see us; we may get a few tips on how to improve our social behavior. Our first attempt at explaining terrestrial culture is contained in the recordings aboard the Voyager probes (Chapter 8). Exchange of such a culture capsule could be a good way of initiating relations with aliens.

But there are more serious problems in the long-term aspects of contact with extraterrestrials. Will contact lead to trade? Or war? Or integration? Any form of trade, apart from swapping souvenirs, seems unlikely because of the long distances between stars; freight rates would be too high. War is possible if we perceive some deliberate threat, or if our visitors willfully ignore our legitimate rights. Living things of all kinds on this planet regard territorial rights particularly dearly; visitors would realize that violating territorial rights is asking for trouble. They would do well to make it clear that they are purely tourists, on a temporary visit.

Can two widely dissimilar societies come to a stable and lasting relationship? This is a question which I do not believe it is possible to answer without knowledge of contacts elsewhere in space. On Earth, the evidence is depressing: advanced societies have always swamped the less fortunate. In many cases, such as with the American Indian, cultures have been destroyed and new forms of government imposed. I do not relish the prospect of our becoming a mere outpost of a galactic empire, ruled from afar like the nations conquered by Rome. We must make it clear from the start that while we are happy to engage in a limited cultural and intellectual exchange, we do not wish to be colonized—and then hope. Because, in the final analysis, any contacted planet will be at the mercy of its visitors.

Although I have looked at the problem so far as though we were on the receiving end of a visitation, the truth of the matter is that if we do not find evidence of advanced extraterrestrials despite serious efforts at picking up messages from the stars, then first contact will probably be the other way around. As we have seen, in only a few centuries starships could be leaving Earth for Alpha Centauri and all points beyond. In the near future, therefore, *we* will be the Little Green Men entering other people's skies. Looking at the problem from this perspective gives a better idea of how visitors ought to behave, and what a civilization's response to them might be.

First, we should examine our moral obligations toward any extraterrestrials. Perhaps our primary principle should be that we not attempt to contaminate them with any of our preconceived notions: there should be no interstellar evangelism; we must accept them as they are. Many cultures on Earth have been destroyed by attempts to "civilize" the natives to Western standards.

Robert A. Freitas has discussed the basic principles of metalaw, the legal rules which we should apply to our dealings with alien races. Both contactor and contacted will benefit from such a procedural outline. The need to guarantee physical security is the first of these requirements. A second is that we should be careful of contacting another culture that might be unable to cope with the shock of meeting us—the so-called principle of noninterference. Above all, there is the golden rule, or the theory of reciprocity: treat others as you would want them to treat you. But is this such a good idea as it seems? Andrew G. Haley, widely regarded as the father of metalaw, thinks not, because it implies that our wishes would also be those of aliens, which is not necessarily so. Instead, we must take their wishes into account. So, Haley says, the rule should be: do unto others as they would have you do unto them.

Austrian legalist Ernst Fasan has formulated eleven rules of metalaw to regulate conduct between aliens. The rules state the rights and obligations of civilizations, including the doctrines that no partners may demand or be required to perform impossible or harmful acts, that civilizations have the rights to their own living space and self-determination, that preservation has precedence over development, and that all societies have the right to defend themselves against the harmful acts of others.

Anthropologist Barbra Moskowitz has also been concerned with our moral obligations toward extraterrestrials. She notes it is essential to practice the doctrine of free choice: that any society must have the right to freely accept or reject any suggestions or advances we may make to them. We must accept their decisions as final. We must observe any civilizations first before we make physical contact with them, so that we know the best way to deal with them and what their reaction to us is likely to be. She terms anthropologists "cultural interpreters" who will help explain our intentions to the visited group, and defend the aliens by ensuring that we adequately understand their viewpoint. If they consider us a threat, she says, they will send a war leader, or if they consider us a supernatural occurrence they will send a religious leader. "We must accept the leadership of those designated to deal with us."

So how do we go about establishing contact? I believe that interstellar signaling not only is the best way of initiating relations with other technical civilizations, but also that it is the most likely, because unless we find some way of cracking the light barrier it will always be easier to send and receive signals than to travel physically between stars. This exchange of messages, which will take centuries or millennia, will give both sides ample time to size each other up. Exchanges at first should be restricted to intellectual matters, such as science and mathematics, but eventually cultural and physical details must follow—these will be both the most difficult to understand and in many ways the most disturbing. We could well find ourselves talking to a computer.

With the directives of metalaw in mind, only when both sides clearly indicate that they wish to meet one another should there be any physical contact. Therefore, long-range interstellar travel should be undertaken only as an adjunct of interstellar signaling. Anyone out there may be standing off from Earth until they get a come-on signal from us. If so, then they have the same attitude to metalaw as we do.

But of course there are likely to be many inhabited planets around the Galaxy whose citizens do not have the technological ability either to receive our signals or to tell us to come on down. What should we do if we discover these on our interstellar explorations? The principle of noninterference says we should tiptoe away again, for even the most passing contact with aliens is certain to change the entire development of a primitive race. Yet scientific curiosity is sure to be aroused by such a discovery. The social anthropologists will have a field day—new forms of art and expression, new forms of government and personal freedom, all providing a perspective on our own way of life Perhaps a limited and circumspect survey-cum-contact will be made, but if so, it is almost certain to leave no trace. Therefore, if Earth had been visited in the past, we should not expect to find any sign of it.

For our own safety, we might not want to make personal contact with every new species that we discover. One suggestion, by anthropologist Shirley Ann Varughese, is that we should grade alien societies according to their levels of social and technical development. For instance, we can compare the time it takes for a civilization to reach certain stages of technical development (the technical quotient) with the time it took to reach the same stage on Earth; we can do the same thing for social development (the social quotient). A comparison of their technical and social quotients will give an assessment of their culture. We might choose to steer clear of a society with a high technical quotient (i.e., rapid technical advance) and low social quotient (i.e., slow social advance); but they will probably be on the way to self-destruction anyway. Fortunately, I believe we will be more likely to find aliens with high social development but low technical development, as in many remote tribes on Earth.

Being the superior civilization in any contact is reassuring, because members of Western civilization, at least, are used to having the upper hand in dealing with people on Earth. We have the luxury of knowing that the maximum freedom of action is ours, and the excitement of knowing that we will have a major influence in what goes on in the Galaxy in the future. But such a situation will happen only if advanced technical life is rare in the Galaxy. We may indeed turn out to be masters of all we survey, but we do not yet know that. In fact, the debates about extraterrestrial life and the attempts to detect messages from the stars have assumed the opposite. We may yet find that we are humble in the family of galactic life.

Being on the receiving end of an encounter with life forms more advanced than ourselves I find somewhat unappealing. Certainly there will be many things we can learn from our superiors, even by radio, but there is a danger that we might be told too much, too quickly. Civilization has progressed by the desire to find its own answers; knowledge that ready-made answers are there for the taking may stifle our own development rather than aid it.

If we do pick up a message from the stars, we must decide whether to respond to it, and how. Who will make these decisions? Perhaps they should be debated publicly, like the issue of genetic engineering. Once radio contact has been made, there is the possibility of personal contact to follow. We must decide whether we want this.

Eventually, although I confess it is impossible to see clearly this far into the future, we will be absorbed into a galactic culture, let us hope to the benefit of mankind. In fact, there seems little point in setting up such a system unless it *is* beneficial: galactic conflict will do no one any good (we would surely have seen clear signs of it by now), and an interstellar empire would be too difficult to administer, if only because of the time taken to send directives.

Whichever way around contact occurs, I believe that if extraterrestrials exist, contact with them is inevitable. Whatever our misgivings, we must be ready to face the existence of other beings in space. First contact will be the end of our isolation, and it will also be the end of our innocence.

REFERENCES

Christian, J. L. (ed.), *Extraterrestrial Intelligence: The First Encounter* (Buffalo, N.Y.: Prometheus Books, 1976).

Freitas, R. A., Jr., *Mercury*, March–April 1977, p. 15.

Maruyama, M., and A. Harkins (eds.), *Cultures Beyond the Earth* (New York: Vintage Books, 1975).

PART THREE

VISITORS FROM THE STARS?

11

The Case of the Ancient Astronauts

If there are advanced, starfaring civilizations out in the Galaxy, where are they? The apparently innocuous observation that there are no extraterrestrials on Earth has been turned to make a telling point against those who profess that advanced life is prevalent throughout space. If such beings existed, the argument runs, they would have explored and colonized the Galaxy as we have explored and colonized the Earth.

This line of argument was developed by Michael Hart of the Goddard Space Flight Center in a paper which has caused considerable consternation in the exobiology community, and the gravity of the paradox he points to has been underlined in an independent analysis by David Viewing of the British Interplanetary Society. Hart imagines that we might in the future send colonizing expeditions to all the nearest likely stars—say, those within 20 light-years. We would be hopping into space like the Polynesians spreading from island to island in the Pacific. Each of our colonies would eventually send out their own expeditions, which would in turn give rise to more, and so on, creating an ever-increasing cascade of humanity that would spread across the Galaxy like ink over blotting paper.

Assuming that we used starships capable of traveling at one-tenth the velocity of light, as the Daedalus probe described in Chapter 6 is designed to do, the frontier of space exploration would advance on a sphere whose radius increased at one-tenth the speed of light. At that rate, most of our Galaxy would be colonized in 650,000 years. Even if we slowed things down by assuming that the time between voyages is roughly the same as the duration of a single voyage, say 50 years, then the Galaxy would still be spanned in less than 2 million years.

The spread of a civilization into the Galaxy has been modeled on a computer by Eric M. Jones of the University of California's Los Alamos Laboratory. Built into the model are additional controls on population growth and levels of immigration and emigration that a civilization might impose. Even with these restrictions, the results showed that the Galaxy would still be colonized in 5 million years. This seems to indicate that, while it will not take us long to spread throughout the Galaxy, we must be the first (or almost the first) to try it.

A further hint that we may be near the forefront of galactic development comes in a study by computer scientist James R. Wertz, who divided the growth and spread of civilization across the Galaxy into three stages: a slow evolutionary phase; a rapid colonization era; and a period of slow growth or decline as the Galaxy becomes fully inhabited. His suggestion is that the Galaxy is now near the transition between stages one and two. If so, we could soon be leading the spread of life into space.

To the question Where is everyone? Wertz answers: "Everybody is still at home, looking at the stars and wondering where *we* are." But N. H. Langton of Robert Gordon's Institute of Technology in Aberdeen is one who feels driven to the dismal conclusion that there can be no other civilizations in the Galaxy at present. The reason, he believes, is that life is so unlikely to reach an advanced stage on two planets at the same time that technological civilizations never coexist; each will have died out by the time the next arises.

It seems to be a widespread assumption that most civilizations have a limited life span; that they live for perhaps a few thousand or a few million years. Perhaps some reach a plateau where they abandon science and technology in favor of sitting at home uncommunicatively to meditate (or vegetate). Others die out altogether. I find this gloomy assumption difficult to accept. Even the dinosaurs reigned supreme for hundreds of millions of years. Why should an advanced civilization die out? What kills it off? This almost passive acceptance of a cosmic

Armageddon seems to deny the natural instinct for self-preservation. Once a civilization has begun to colonize space, does this not insure the civilization against localized catastrophes? On the face of it, there seems no reason why humanity should not exist forever. Or is there some fatal flaw in evolution that has already condemned us irrevocably to eventual extinction? Biologist Peter Molton has spoken of what he terms nature's joke on mankind: given no new frontiers, the aggressiveness that gave us civilization will turn inward to violence and war, a tendency which is perhaps already becoming visible in society today.

As discussed in Chapter 1, a long average lifetime for civilizations implies that there should be many civilizations around at the moment, and therefore numerous potential interstellar travelers. One line of explanation frequently offered for the absence of extraterrestrials on Earth is that they have chosen not to come here for some sociological reason: either they are keeping Earth as a space wilderness area or zoo, or they have no interest in space exploration, or their civilization was wiped out before they reached us. But all such attempted explanations must apply to *every* civilization *all* the time—an implausible uniformity in a Universe of such variety. Thus we are led into what we might term Hart's paradox: one cannot have abundant extraterrestrials without abundant evidence for them on Earth.

While the paradox doesn't rule out the existence of simpler forms of life on other planets, it does imply that the factors in the Drake equation governing emergence of advanced technological life have been overestimated. But the conclusion that we are alone or among the first advanced civilizations to arise in the Galaxy is not yet inescapable: a paradox can reveal as well as refute. The famous paradoxes of Zeno and Olbers were resolved by factors not even dimly perceived at the time they were advanced. What Hart's paradox may reveal is that our perception of the problem is far too naïve.

For instance, even if there are one million civilizations like us or more advanced throughout the Galaxy, as the standard calculations suggest, the nearest would still be several hundred light-years away. They may never reach here in manned craft because the time involved dissuades them from journeying in person beyond a few tens of light-years of home. There may of course be exceptions to this rule, but they will be affected by other factors described below. In any case, interstellar travel is certainly more difficult than sending radio messages.

Therefore it should still be worth searching for messages from the stars, even though the chance of a personal visit from the extraterrestrials is very slim.

Would civilizations indiscriminately send manned expeditions to colonize the stars as Hart supposes? The cost and use of materials for space purposes must be justified by any civilization against competition from its other needs. Most likely, the first stages of interstellar exploration will involve unmanned pathfinder probes, like Daedalus or Bracewell's messengers. Only then would manned missions follow, one at a time, to selected targets. Each new colony would not send out its own explorers until it too had scouted carefully ahead with unmanned probes—a survey which might take 1,000 years or more. At this rate, it would take many times longer than the age of the Galaxy to complete the spread of colonizers, assuming they wanted to saturate the Galaxy, which I doubt.

I presume that any manned interstellar expeditions will not be colonists but observers, like the crew of the U.S.S. *Enterprise* in TV's famous *Star Trek* series. Any colonies that are set up may only be temporary, with the starfarers choosing not to despoil but instead moving on like nomads of the stars.

There may be strong effects of natural selection which mean that nature prefers nomads to habitual colonizers: civilizations that do not realize the importance of severely limiting their own numbers soon suffer a population crash, and possibly extinction. Occasionally, civilizations will be forced to move on because of the impending demise of their own star. But there is no reason why they should choose more than a handful of nearby sites to move to, around stars with long-term prospects.

Expansionism need not be the only philosophy for technological life. The lemming-like rush to expand that is current among mankind today is a result of our relatively primitive stage of technological development, although already there are signs of a change. In general, living things on Earth are in a balance with their environment, as indeed were the hunting bands of early men on the plains of Africa. A better clue to the true nature of spacefarers may be the stable populations of bushmen and nomadic tribes. Even a civilization that grows only slowly will eventually have to stop once it has colonized the entire

Galaxy. But long before that stage is reached it will have made its decision to stabilize.

Finding a suitable new world and setting up a colony there may be far more difficult than we suppose, particularly since the most enticing planets are already likely to be inhabited. In search of a solution to Hart's paradox, astronomer Laurence Cox of the Hatfield Polytechnic Observatory, Hertfordshire, England, has pointed out that our bodies need twenty specific amino acids, not all of which might be available on other planets. Hart's argument assumes that our biochemistry is typical of living things throughout the Galaxy, but such a requirement would by itself severely limit the number of likely civilizations in space. Unless there is some reason why our biochemical makeup should be favored, alien plants or animals will most likely have little food value to us, or will be poisonous. Another problem is the danger of disease from microorganisms in air or water. Even subtle matters such as the percentage of oxygen in the atmosphere would affect our physical efficiency. The product of billions of years of evolution on one planet is entitled to feel ill at ease if transferred to an alien biosphere. Establishing artificial biospheres on sterile planets or in space arks would be a solution, but would add greatly to the difficulty of setting up home around a new star.

In a direct answer to Hart's paradox, Cox notes that a population growing at our current rate would soon outstrip the rate at which it could colonize new planets. For us to populate one planet around each star in the Galaxy to the current density on Earth in the 650,000 years envisaged by Hart would require that our population double in not less than 20,000 years, as against 30 years now. Cox concludes that by the time any society reaches the stage of interstellar travel it will already have had to stabilize its population, and the paradox, which assumes continued expansion, is avoided.

Further considerations have been voiced by David Stephenson of the Institute of Space and Atmospheric Studies of the University of Saskatchewan, who notes that the time scale of millions of years required to colonize the Galaxy is also the time scale of evolution. Over the course of galactic colonization, one would expect strong evolutionary effects to become noticeable; think, for instance, of the evolutionary changes in man over the past few million years. It is probable, says Stephenson, that the inhabitants of a starfaring vessel will be adapted to deep space travel. The resulting creatures will be physically very different from ourselves, and would actually find planetary living uncomfortable or even hostile. Although the space creatures may pay temporary visits, they would be unlikely to colonize.

Furthermore, the limited space and consumables on interstellar ships means that such travelers would need total control over reproduction; they would, in fact, be conditioned to living in an environment of zero population growth. Starfarers will not only be physically different from us, but also culturally so: their motivation is purely one of curiosity, not of expansion and demand for new resources that has so far governed human spread on Earth.

As elaborated in previous chapters, it is more likely that intelligent machines will undertake interstellar exploration, rather than humans. Such an intelligent visitor could be living in the solar system now, and, being a machine, it would survive happily in space without the need to colonize a planet. Perhaps the whole Galaxy has by now been scouted by intelligent machines and only in a few instances have biological creatures followed.

My conclusion, then, is optimistic. I do not think that the absence of extraterrestrials on Earth today is proof of their total absence in the Galaxy. Rather, I would interpret it as supporting evidence that civilizations stabilize their growth and do not colonize.

Conceivably we have been visited once or many times in the past by interstellar nomads, none of whom chose to stop. If, by chance or design, we were visited relatively recently, signs of the visit might still be evident in artifacts and legends. Some writers claim that such evidence exists. Others maintain that the extraterrestrials are here today, and that we see them (or their craft) in the form of UFOs. This means that we have tangible messages of other civilizations from the stars. We shall examine these popular assumptions in this and following chapters.

The best-known evangelist for the belief in visits from ancient astronauts is Swiss author Erich von Daniken, whose books on the subject have consistently outsold all other authors on the same theme. Von Daniken's thesis, nowhere better exposed than at the end of his fourth book, *In Search of Ancient Gods*, is that the emergence of true man, genus *Homo*, from the apes occurred with the help of genetic manipulation by aliens. He speaks of a "sudden" advance in intelligence.

Actually, fossil evidence reveals that the earliest known members of *Homo* lived at least 3 million years ago. Yet, according to von Daniken, earthlings of only a few thousand years ago were still so dim that they needed alien help to build the Egyptian pyramids and Inca cities. Apparently another round of genetic improvement was called for at that time. But there is no evidence of any changes in man's physical or mental capabilities for at least 10,000 years, since the time of the Neolithic revolution; the only changes have been cultural ones. Even put at its clearest, von Daniken's theory is confusing and contradictory.

Von Daniken's success is due not to any originality of evidence, since all the authors in this field cite much the same examples (mostly quoting each other). Rather, it is due to his highly effective style of writing (or rather that of his rewrite man, Wilhelm Roggersdorf). One feels that his tendentious style would have made him a fearsome advocate in court had he taken to the legal profession.* And it is on a courtroom analogy that his work is best judged. When all the evidence is taken into account, does he prove his case beyond a reasonable doubt?

Avid von Daniken readers will have spotted many examples of his persuasive style with which, without saying anything wrong or making a definitive statement, he nevertheless draws you, the jury, to the conclusion he wants—usually by means of the rhetorical question. I shall cite one example, from *Chariots of the Gods:*

> Various people knew the technique of embalming corpses, and archaeological finds favor the supposition that prehistoric beings believed in return to a second life, i.e. a corporeal return. Drawings and sagas actually indicated that the "gods" promised to return from the stars in order to awaken the well-preserved bodies to new life. Who put the idea of corporeal rebirth into the heads of the heathen? And whence came the first audacious idea that the cells of the body had to be preserved so that the corpse, preserved in a very secure place, could be awakened to new life after thousands of years?

Overlooking the pejorative term "heathen" here, let's look at the technique used by the Egyptians for mummification. First, the brain

* Alas, this talent did not prevent his conviction in 1970 for embezzlement, fraud, and forgery.

was pulled out through the nostrils by tweezers. Then the skull was filled with resin. The vital organs were removed, and the body was coated with embalming fluid which solidified so that it would later have to be chipped away by hammer and chisel, or heated to 500°C to melt. To borrow von Daniken's rhetorical technique: Does this sound like the preparation for the reawakening of the dead?

There are many examples of religious drawings and stylized wall paintings or sculptures shown by von Daniken that seem to depict astronauts in spacecraft or spacesuits. Of all these, perhaps the best known is the so-called Great Martian God, a cave drawing discovered in 1956 by the Frenchman Henri Lhote at Tassili in the Sahara. Lhote described the figure as dressed in a ritual mask and costume, part of an artistic tradition known as the period of the round heads. But in the "ancient astronaut" books the expert witness is not called to give evidence that puts the drawing into context, and it becomes a mysterious portrayal that can only be of a man in a space helmet.

Perhaps the "Martian god" is not very convincing; neither is the "Palenque astronaut," an engraving on the lid of a Mayan tomb said by von Daniken to resemble an astronaut but actually a ceremonial depiction of the Mayan king Pacal, whose bones the tomb contained. Altogether more impressive is the famous Piri Re'is map. This is a map of the Atlantic Ocean and its surrounding continents, which belonged to the sixteenth-century Turkish admiral Piri Re'is, and was discovered in the Topkapu Palace, Istanbul, in 1929. Von Daniken claims that the coasts of North and South America and even the contours of the Antarctic are "precisely delineated" on this map. His theory: that the map was drawn from aerial photographs taken by a spaceship hovering above Cairo.

The truth is a little more mundane. According to expert cartographer Charles Hapgood, large sections of the South American coastline on the Piri Re'is map are missing. Hapgood does not have a very high opinion of the accuracy of the map, and certainly does not suggest it was made from aerial photographs, although von Daniken gives exactly the opposite impression in *Chariots of the Gods*. The moral here is: always check with the original source.

Careful comparison of this exhibit with modern maps reveals other embarrassing discrepancies. Cuba is misshapen and misplaced; it is impossible to confidently identify the Amazon or the Río de la Plata;

and Antarctica does not even appear. Instead, the coast of South America meanders across the bottom of the map as though the cartographer had lost his bearings. For the result of a space-borne survey, it is feeble—although as an example of the work of early navigators it is impressive.

In fact, the Piri Re'is map looks remarkably like other charts of the time known as portolanos, which were made from compass bearings and distance estimates. As on other portolanos, the Piri Re'is chart contains two mother compasses, plus several smaller ones, for picking off magnetic headings. The chart naturally becomes distorted down the coast of South America because of the increasing angle of divergence between the magnetic and geographic poles and the mariner's uncertainty of his actual distance traveled. The Piri Re'is chart is dated 1513, and is a compilation of several earlier maps going back to Columbus. In 1569 Gerhardus Mercator invented his famous Mercator projection, and considerably better charts were soon being produced. But no one suggests that these were the work of extraterrestrial visitors.

The ancient "batteries" found near Baghdad that von Daniken marvels over may be examples of very primitive electrical cells, possibly used for electroplating at about the time of Christ. A replica was tested in 1960 and produced half a volt of electricity for eighteen days—impressive for 2,000 years ago, but hardly of use to advanced space voyagers.

The lens from 700 B.C. in the British Museum which von Daniken says in *Chariots of the Gods* needed "a highly sophisticated mathematical formula" to grind is in fact a piece of natural crystal polished around the edges.

The rustproof pillar near Delhi which von Daniken says is made of "welded parts" and of an "unknown alloy" is actually a single piece of pure iron—the result of advanced, but nevertheless terrestrial, metallurgy of about A.D. 500. In an interview with *Playboy* magazine in 1974, von Daniken admitted that new investigations had made him change his mind, "so we can forget about this iron thing."

Von Daniken's vivid imagination plays freely over Mayan drawings and carvings. As he so rightly says in *In Search of Ancient Gods*: "So far only a minimum number of the Mayan picture writings has been deciphered, so there is plenty of scope for my assumptions." This is the whole problem: too much fantasy, too few facts. Never has von Daniken given the game away so clearly.

Von Daniken makes accurate statements and then contradicts them, as in this example from Chapter 6 of *Gold of the Gods*: "The oldest demonstrable remains of forms of life on Earth were discovered in 3.5-billion-year-old sedimentary rock in the Transvaal, South Africa [true]. Their stage of development corresponds to that of the blue algae living today [true]. But 3.5 billion years ago there was no kind of organic life on our planet" [contradiction].

In *Chariots of the Gods* he notes correctly that the pyramids were apparently aligned on astronomical objects and that the Egyptians had an accurate calendar based on the rising of the sky's brightest star, Sirius. Then he claims in contradiction that we have "very little evidence of an early Egyptian astronomy" and wonders why they were interested in Sirius! (For another hypothesis concerning Sirius, see Chapter 13.)

He does not suggest extraterrestrial influence on the building of Stonehenge and other megalithic monuments which are aligned astronomically and actually predate the pyramids; evidently he accepts that Europeans were advanced enough to understand astronomy, while the Egyptians were not. This highlights a curious bias in von Daniken's world coverage: European examples are missing. They're too familiar. The omission is strange, for the achievements of European cultures, as for instance documented by Colin Renfrew in his book *Before Civilization*, were in many cases more advanced for their time than any of the examples von Daniken quotes.

In *Chariots of the Gods* von Daniken scoffs at archaeologists who claim that the pyramid builders pulled their stone blocks by ropes over wooden rollers, both of which he asserts were "nonexistent." But friezes of the period show . . . pyramid builders pulling stone blocks over wooden rollers. Ropes have been found in the quarries from which the pyramid stones came, and the Egyptian trade in cedar logs from Lebanon is well documented.

Von Daniken overlooks the fact that the building of the pyramids was no overnight affair, but was a slow process of development over many centuries. The first pyramid, that of Zoser, was built in a step shape. The second was built at too steep an angle and collapsed, so the third, which was under construction at the time, had its angle of slope changed, producing the celebrated Bent Pyramid. Such disastrous mistakes do not seem the work of highly advanced "gods" from space.

Skipping continents and coming 3,000 years forward in time, von Daniken draws attention to the precision jointing of stone walls built by the Incas in South America. But equally fine stonework, such as the Colosseum in Rome or the Parthenon in Athens, was being undertaken in Europe over a thousand years before the Inca cities were built. The medieval cathedrals of Europe with their astounding stonework are contemporary with Inca civilization, and what's more, they contain soaring arches, which von Daniken's "gods" never taught the Incas. Von Daniken omits dates, confuses cultures, and deliberately underestimates the abilities of ancient peoples, particularly those who seem strange and remote to us.

Another example of von Daniken's cavalier treatment of the facts concerns the markings on the plain of Nazca in Peru. The marks have been made simply by removing the brown stony surface of the plain to reveal the lighter-colored soil underneath. Though no one (except von Daniken) seems quite sure of their purpose, archaeologist Maria Reiche, who has made a lifelong study of them, is convinced that astronomical alignments, like those of the earlier European megaliths, are at least partly involved. Many of the figures drawn on the plains are representations of animals, apparently portrayals of constellation figures that played important parts in Inca mythology. Yet von Daniken interprets the markings as aircraft runways and landing bays. Why advanced spacemen with technology far in advance of that which set our own lunar module on the Moon should need gigantic markers and aircraft runways eight miles long is left unexplained. But what is clear is that no aircraft, or any other craft, ever set down there. Had it done so, it would have left its own indelible markings in the loose desert surface, like the tracks left by vehicle tires today.

Perhaps von Daniken's most outrageous misrepresentation is of the work of Thor Heyerdahl on the Easter Island statues. To von Daniken, these gigantic statues carved out of volcanic rock could only have been the work of spacemen, blatantly ignoring the fact that Heyerdahl had filmed the entire process of cutting and erecting a statue, which proved that it could be done by the existing population of the island without outside help. Heyerdahl long remained silent, in the belief that, as he says, "anyone stupid enough to take this kind of hoax seriously deserves to be cheated." But in a recent critique of the works of Erich von Daniken by Ronald Story, *The Space Gods Revealed*, he published an open attack on what he termed the "entertaining fiction" of von Daniken: "The general reader who cares to know has the right to be informed that there is not the slightest base of fact in what von Daniken writes concerning the origin of the giant statues on Easter Island. We know exactly how they were carved, where they were carved, why they were carved, and when they were carved. Together with my colleagues I am to blame for not promptly having used the modern mass media for telling the public not to take his references to Easter Island seriously."

Jetting around the world in search of archaeological mysteries, or scampering after celebrities to snatch a tape-recorded interview, von Daniken presents an engaging naïveté that has undoubtedly contributed considerably to his popular success. His audience, no more able to test his theories than he is, doubtless admire his brazenness in stepping so firmly on so many toes—and there is, of course, an overwhelming reaction that anything expressed so confidently in print *must* be right. But, in omitting so much of the evidence, he does not give his readers a chance to judge for themselves how accurate or plausible his assertions are. Unless his audience are willing to become experts on archaeology and astronomy, they must take von Daniken's views on trust: counsel for the prosecution, without one for the defense.

Von Daniken's most celebrated case is that of the mysterious underground tunnels with glazed walls in Ecuador containing golden treasure, which he describes in *The Gold of the Gods*. Von Daniken vividly recounts his descent into the caves, led by his Ecuadorian guide Juan Moricz, who, says von Daniken, stopped him from taking photographs. Therefore we never actually see the supposed treasure hoard; and the pictures of the tunnels reproduced in the book are apparently someone else's. But this does not seem to matter, because von Daniken next shows photographs of this "Inca treasure" in the possession of Father Carlo Crespi, a Catholic priest of the Church of Maria Auxiliadora in Cuenca. According to von Daniken, the Indians bring to Father Crespi the most valuable gold, silver, and metal objects from their hiding places, and have done so for decades. These ought to be of the greatest archaeological importance, for they contain a previously unknown Inca script. So what's wrong here?

The German news magazine *Der Spiegel* tracked down Juan Moricz, the Ecuadorian who, von Daniken claimed, had personally conducted him through the caves on his purported visit so vividly described. But Moricz denied that von Daniken had ever been in the caves.* Instead, Moricz said, he had only taken von Daniken to a blocked side entrance. There Moricz was pumped for all the information he had, which von Daniken later passed off in loving detail as his own experiences. Von Daniken admitted his lie on a BBC television documentary in the Horizon series, broadcast in November 1977 (shown in the United States on the PBS *Nova* series). He said that his invention of the story was "not important," and appealed to artistic license. Yet this story is contained in a book which begins: "It could easily have come straight from the realms of science fiction if I had not seen and photographed the incredible truth in person."

Of course, if von Daniken had real evidence he would not need to hoodwink the public.

I have been in similar caves: gigantic halls that stretch far underground, with shiny walls so smooth that they could never have been polished by the primitive men who lived in them. They have an eerie ventilation system that keeps the temperature remarkably constant all year round, like the controlled environment of an underground shelter. Strange, glazed shapes decorate the walls, ceiling, and floor as though carved from the very rock by processes that we could not imitate with our technology today.

The caves are at Cheddar Gorge and Wookey Hole in England. They were formed by the natural action of water, as were the glazed walls and the stalactites and stalagmites that adorn them. According to the BBC Horizon program, a British expedition to the Ecuadorian caves in 1976 found them to be natural in origin, with no sign of any gold valuables.

Archaeologist Pino Turolla of Miami, Florida, has reported that the "gold treasure" of Father Crespi is crude metalwork made from tin and brass by the local Ecuadorian natives, who trade it to him in return for food and clothes. Crespi, although lovable, is a somewhat eccentric

* The exposé was featured on the front cover of *Der Spiegel* as "Der Daniken Schwindel," which scarcely needs translation.

and unreliable old man. Turolla says he has seen a copper toilet-tank float among Crespi's "gold of the gods."

The "Inca writings" on these cheap imitations are present-day doodles. Von Daniken reproduces in his book one embossed metal plate showing an Egyptian-style pyramid attended by two elephants, which, as he rightly says, "the artists could not possibly have seen in South America around 12,000 B.C." As with the KLEE-TV hoax reported in Chapter 11, such a glaring anomaly should immediately have revealed the truth. But instead, it is presented as an added "puzzle" to impress the credulous.

Von Daniken, who cheerfully describes himself as a "Sunday archaeologist," presented further exhibits from Peru on a BBC television program, *Tonight*, in 1976. He showed slides of small carved stones, apparently predating the Incas, which depict advanced surgical knowledge, including heart transplants and blood transfusions, being performed under anesthesia. The museum at Ica in Peru contains thousands of these stones, which, surprisingly for such valuable archaeological specimens, are also on sale in unending supply to tourists.

The BBC Horizon investigation the following year cleared up the mystery. These, again, are modern artifacts made by the locals. A stoneworker admitted he had made all the carvings, taking his ideas from newspapers and magazines. He carved a "heart transplant" design for the cameras. The dark, weathered look of the stones is obtained by baking them in donkey dung and blacking them with boot polish. The Horizon team took a stone from the museum along with their specially made "heart transplant" stone for analysis at the Institute of Geological Sciences in London. Scientists there confirmed that the sharpness of the etched grooves and the lack of weathering on the "real" stone from the Ica museum meant that it had been carved recently, and was the same as the fake stone carved for the cameras.

It seems pointless to continue.

"Von Daniken's books are not written to persuade the informed reader. They are a romanticist's fiction," notes Stanford University radio astronomer Ronald Bracewell. "Von Daniken's books tell us not about the ostensible subject matter but about the society which buys them so eagerly. For those who need certainty and stability in a world of jolting change and ambiguity, the attractions of von Daniken's books are that they are understandable and are presented vividly with firm conviction. They represent a substitution of faith for reason. Unfortunately, they can offer only a brief haven as they are so vulnerable to criticism."

One of the major problems faced by exobiologists is to understand how creatures became highly intelligent and technologically advanced. Ascribing it to the influence of ancient astronauts is no solution; that simply transfers the problem elsewhere. The "gods from space" approach hinders understanding of human development rather than helps it.

One point is clear. The stonework of South America, the Easter Island statues, the pyramids, or even Stonehenge, while impressive structures, are all state-of-the-art technology for the human civilizations that built them. Visiting spacemen or "gods" do not work in stone. They use metals and plastics.

For an extraterrestrial artifact to be convincing, it must be clearly out of place—for instance, a transistor radio embedded in a block of clear plastic which examination would show was beyond even twentieth-century technology. The existence of remarkable stone structures in itself proves nothing about the existence of extraterrestrials, but these structures could act as protection or camouflage for the true artifact, such as that transistor radio, which might lie in some as yet undiscovered burial chamber. Alternatively, the structures themselves might act as a disguised pointer to something that lies beyond Earth.

And it is just such a suggestion that we turn to next.

REFERENCES

Bracewell, R. N., *The Galactic Club* (San Francisco: W. H. Freeman, 1975).

Cox, L. J., *Quarterly Journal of the Royal Astronomical Society*, Vol. 17 (1976), p. 201.

Hart, M. H., *Quarterly Journal of the Royal Astronomical Society*, Vol. 16 (1975), p. 128.

Jones, E. M., *Icarus*, Vol. 28 (1976), p. 421.

Langton, N. H., *Journal of the British Interplanetary Society*, Vol. 29 (1976), p. 465.

Molton, P. R., *Journal of the British Interplanetary Society*, 1978 (in press).

Stephenson, D. G., *Journal of the British Interplanetary Society*, Vol. 30 (1977), p. 105.

Story, R. D., *The Space Gods Revealed* (New York: Harper & Row, 1976).

Viewing, D. R. J., *Journal of the British Interplanetary Society*, Vol. 28 (1975), p. 735.

Wertz, J. R., *Journal of the British Interplanetary Society*, Vol. 29 (1976), p. 445.

12

Signpost to Mars

In the stark light of an alien star, a space transporter hoves to above a small planet. Sunlight flashes on vast sheets of blue-green sea, while wreaths of dazzling white clouds obscure large areas of land. There is liquid water on the planet in abundance. The spaceship's instruments reveal free oxygen in the air—a sure sign of photosynthetic plants. There is life on this planet!

The eager crew see no lights burning on the night side of the planet, and the spaceship's radar detects no artificial satellites in orbit. No radio transmissions are discernible, and infrared scanners fail to find any hot spots that would be the telltale signs of industry. This planet does not contain technological life.

But telescopic inspection reveals a checkerboard pattern—fields under cultivation by intelligent beings! An automatic landing probe is dispatched to send back data and photographs from the planet's surface. Its pictures are sensational. Peering at their monitor screens in the mother ship above, the space adventurers see strange, tall plants growing from the rich soil of the planet. Its air is dense, warm, and moist. Other lander probes show animals scurrying on the surface. The planet teems with the most remarkable variety of life. A landing party is dispatched.

Cautiously, they drop toward an area flanking a major river with a fertile flood plain. Inspection from orbit has revealed large and puzzling structures here, throwing long, pointed shadows under low illumination. The explorers find . . . an advanced, socially organized people, on the verge of the technological age.

An encounter like this seems certain to happen sooner or later during the colonization of the Galaxy. But what would explorers do if they came across a society of intelligent beings that seemed likely to reach the stage of high technology in the next few thousand years? A direct approach—"Hello, we're from space"—would be little good. The inhabitants would either recoil at the thought of men from space or worship them as gods. In either case, a meaningful exchange of views would seem unlikely. The visitors would need to adopt a more cautious approach, probably concealing their origin and their motives for making contact.

Direct supply of technological knowledge and artifacts would also be fruitless, for the technologically unsophisticated inhabitants would be unable to make use of them. To quote an analogy by the American physicist Philip Morrison, it would be as though an angel had written $E = mc^2$ in the notebook of a Renaissance mathematician. Although technological aid out of context would therefore be no good, help with existing technology is a possible approach.

What the visitors might like to do is leave a store of information that will take on meaning when the civilization comes of age technologically. But leaving it on the planet might not be the best place, because of the many potential natural or human-made mishaps that could befall it. Equally, the space people could have no way of knowing whether the emergent civilization was to be trusted with such advanced and potentially dangerous knowledge, which might, for instance, include information on new energy sources and hints for interstellar travel. The visitors might therefore place a databank somewhere off the planet, leaving behind a subtle pointer to its whereabouts. If the civilization turned out to be destructive, they would eradicate the pointer, thus losing all hopes of finding the databank. Only a conservationist, space-oriented society would preserve the pointer, decode its message, and find the treasure that awaits.

What form would the pointer take? According to Mike Saunders, an English electronics engineer, it might well be a giant stone monument of the sort found in several countries. And the greatest example of all is the Great Pyramid of Cheops, the largest precision-built monument in the world. Does this hold the secrets of a former extraterrestrial contact with Earth?

On Saunders' reasoning, use of a massive stone monument as a pointer has several clear advantages. First, it will not arouse unnecessary suspicion, because it is clearly state-of-the-art technology for the people who built it; if the inhabitants were not already building pyramids when the extraterrestrials landed, they could easily be taught to do so using available techniques. Secondly, a pyramid is so massive that it would last for thousands of years.

Saunders does not deny that the pyramids, or other impressive stone structures, were built by humans. The Great Pyramid at Giza is the major example of an era of pyramid-building whose reasons and techniques are well understood by Egyptologists. One possibility is that space visitors used the building of the pyramids as an opportunity to observe mankind at work, and even to instill in them the virtues of co-operation that they would eventually need to survive. If mankind later took a wrong turn and failed to preserve the works of their ancestors, then they would forfeit this subtle and elegant sign of a visit by extraterrestrials.

Saunders assumed that the massive Great Pyramid, which was built to a precision far greater than any other, might have been designed by the extraterrestrials for construction by the unwitting Egyptians, with certain built-in pointers to an extraterrestrial databank. So are there any signs that a hidden message is contained within the design and dimensions of the Great Pyramid?

It has long been known that the total height of the Great Pyramid, allowing for the destruction of its top cap, is approximately one billionth the distance from Earth to the Sun. Also, scientists have long wondered why the pyramid's fundamental dimensions should incorporate the mathematically significant number π, which is the ratio between the circumference and diameter of a circle, and is roughly equivalent to 3.1416. For instance, the perimeter of the pyramid is 2π times its height, and its slope of nearly 52° is the same as the slope of a triangle whose height is 4 units long and its base π units long. Yet there is no indication that the Egyptians understood the mathematical significance of π. It has been suggested that π became incorporated inadvertently through use of a drum of standard diameter to roll out the pyramid's dimensions.

Saunders' investigation, which included a trip to Egypt to check the pyramid data for himself, led him to suggest that several of the fundamental parameters of the Great Pyramid are pointers to Mars—and, in particular, the strange moons of Mars. In the investigation, those factors of π, 2π, and 4π make a significant reappearance.

Firstly, Saunders notes a connection between what he terms "the most fundamental dimension" of the Great Pyramid, its base length, and the most fundamental dimension of the orbit of Mars, its average distance from the Sun: the average Mars–Sun distance is 1 billion times the Great Pyramid's base length, to an accuracy of about 99 percent. Saunders explains the 1-billion-to-1 ratio (the same approximate ratio as between the pyramid's height and the distance of Earth from the Sun) by the fact that the resulting monument is of imposing size, but still manageable. A pyramid ten times bigger would be impracticable to build, whereas one ten times smaller would be insignificant.

This one coincidence in itself would mean little, but there are several other fundamental factors about Mars to take into consideration. For instance, the Sun is displaced slightly from the center of the orbit of Mars, and the Main Chamber of the Great Pyramid is also displaced from the pyramid's center by the same relative amount. In other words, if the orbit of Mars were drawn to fit within a plan view of the Great Pyramid, the Sun would lie exactly in the Main Chamber, where, according to legend, the Pharaoh was to be laid so that he could unite with the sun-god.

A third suggestive coincidence concerns the access corridor to the Main Chamber. The corridor is offset from the center line of the pyramid, thereby dividing one side of the pyramid into two unequal halves (fig. 1). The ratio between the longer half and the overall length of the side is the same as the ratio between the length of the Earth's year and the length of the Martian year. Therefore, it seems that the three most fundamental parameters of Mars—its orbital size, its orbital focus, and its year—can all be found from the three main features of the Great Pyramid's layout.

There are other indicators to the planet Mars. Most important of these is the slope of the pyramid's north face, which points to a spe-

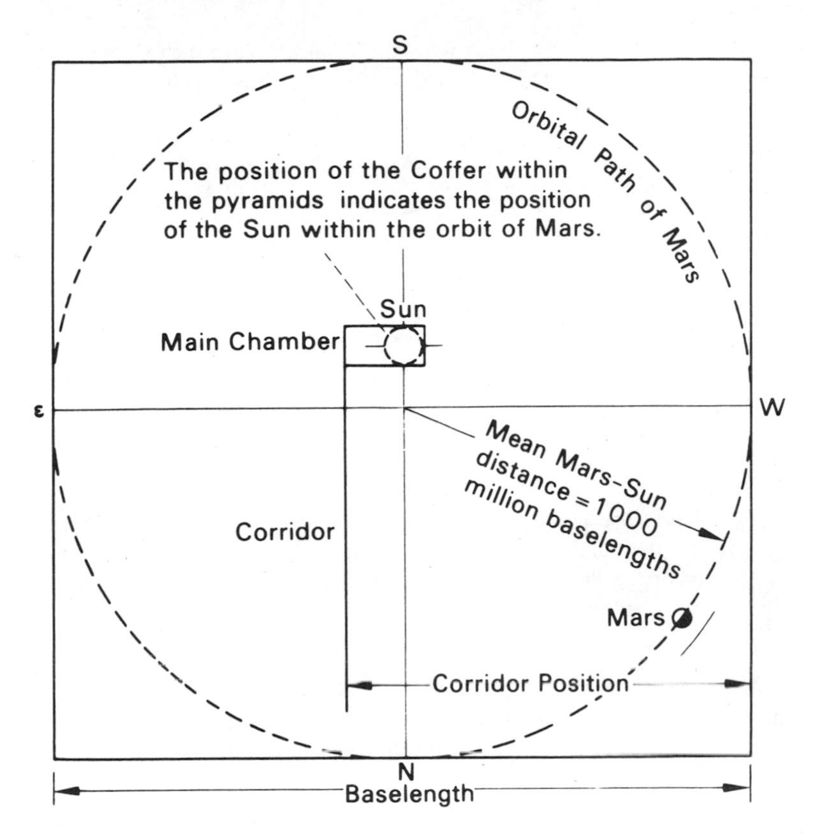

Figure 1. Plan view of the Great Pyramid with the orbit of Mars around the Sun superimposed. Also shown is the access corridor to the Main Chamber that divides the north face of the pyramid into two unequal halves.

cific height above the Earth's equator; it turns out that this height is equivalent to the circumference of Mars (fig. 2).

But what about the actual databank site? This is where Saunders thinks that the hypothetical extraterrestrials may have made cunning use of a naturally occurring coincidence to complete what he terms their "monument test." The coincidence concerns the largest volcanoes on Earth and Mars. The largest volcano on Earth is the island of Hawaii, which lies at a latitude of about 19°N. The largest volcano

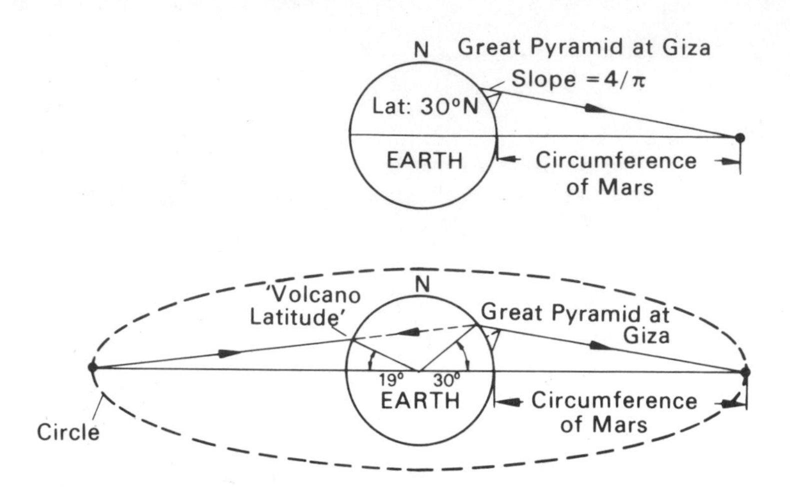

Figure 2. The slope of the Great Pyramid points to a height over the Earth's equator equivalent to the circumference of Mars; a line from this height above the equator drawn through the Earth to the Great Pyramid defines the so-called volcano latitude, 19°N.

on Mars, called Nix Olympica, also lies at a latitude of about 19°N. A line drawn from the Great Pyramid through the Earth to this so-called volcano latitude of 19°N points to a height above the equator which is equivalent to the circumference of Mars (fig. 2). The existence of the volcanoes is not crucial, as such a line would indicate this latitude anyway, but they act as additional pointers to Mars.

The final link in this chain of reasoning is that a line drawn from the orbit of the Martian moon Deimos to latitude 19°N on Mars and continued through to the other side of the planet cuts the surface at the very latitude, 25¼°N, from which a Great Pyramid shape points to the second Martian moon, Phobos (fig. 3). The longitude of the site, 313¾°, is in an area coincidentally named Arabia. In other words, it turns out that the two Martian moons, Phobos and Deimos, are important parts of the indicator system, and that the ultimate target is the larger moon, Phobos.

Actually, the moons of Mars seem a good place for a databank. Our own Moon is too easy to get to. A manned trip to Mars, on the other hand, would need international cooperation of the sort that would

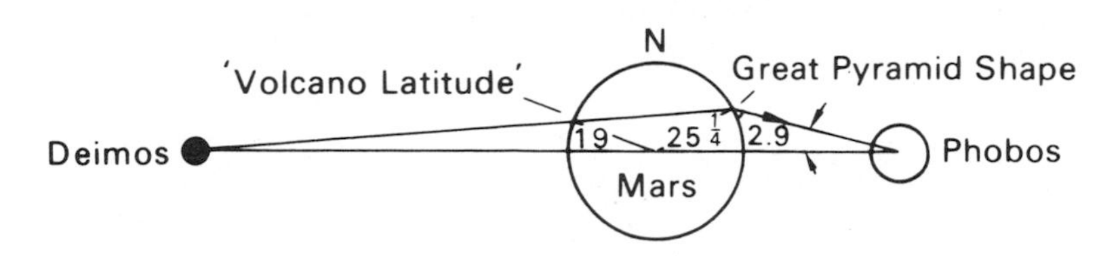

Figure 3. A line from the orbit of Deimos through 19°N on Mars defines the latitude on Mars at which a Great Pyramid shape points to Phobos.

make it impossible for one nation alone to steal the secrets of the databank for themselves. The surface of Mars itself is unsuitable because it is subject to erosion by the planet's atmosphere. But apart from occasional micrometeorites, the surfaces of Deimos or Phobos are erosion-free.

Saunders notes several numerical coincidences which, he presumes, are intended to draw our attention to the moons of Mars. For instance, Phobos orbits Mars π times in the time Earth spins once on its axis—and we have already pointed out the importance of π in the dimensions of the Great Pyramid. Phobos is in fact the only moon in the solar system that orbits faster than its parent planet spins. Deimos, the smaller and more distant of the two moons, goes around Mars $\pi/4$ times for every rotation of Earth—and the slope of the Great Pyramid is $\pi/4$. As another example, the height of Deimos above Mars equals π times the Earth's radius. The relation of the Great Pyramid to the orbits of the Martian satellites is a little too remarkable, Saunders says; he feels that it is difficult to ascribe the affinity purely to chance.

Astronomers know that Phobos and Deimos are strange bodies. Each is a dark, heavily cratered lump of rock, of carbonaceous chondrite composition, similar to that of the asteroid Ceres. Apart from minor irregularities, the two moons are of almost identical shape, like rugby balls, about 27 kilometers and 15 kilometers across at their longest, respectively; they both orbit with their longest axes pointing toward Mars. They may be captured asteroids. Yet their orbits are remarkably orderly: each moves in a nearly circular path almost exactly in line with the Martian equator. Such orderliness is surprising for a capture origin. Alternatively, they could both be fragments of a former, larger Martian moon. But what caused the fragmentation?

And why are Phobos and Deimos the only remnants?

Neither Phobos nor Deimos can remain in its present orbit for long. Tidal interactions with Mars are slowly driving Deimos outward, while Phobos is moving inward. In fact, in about 100 million years Phobos will crash to the surface of Mars. Most astounding of all, close-up photographs of Phobos taken by the Viking orbiter craft in 1976 and 1977 show parallel grooves running around its equator. Saunders' theory presumes that the moons of Mars, while being natural lumps of rock, were deliberately installed in their present orbits by the extraterrestrials to act as part of the indicator system; and one can only speculate at the implications for the theory opened up by the current discoveries about Phobos and Deimos.

What is not speculation is that the slope of a Great Pyramid shape on Mars pointing directly to the center of Phobos cuts the tiny moon's surface at a latitude that can be determined precisely: 12.88°N. A line drawn from Mars to the center of Phobos passes through longitude 0°, the point at which Mars appears directly overhead. Therefore, Saunders believes the extraterrestrial databank to lie at longitude 0°, latitude 12.88°N on Phobos, within an error of about ten meters north or south. On maps of Phobos, drawn from space-probe photographs, the databank site lies a few kilometers to the east of a large crater named Stickney (fig. 4), the maiden name of the wife of Asaph Hall, who discovered the moons in 1877.

What would the databank look like? To minimize the chances of accidental discovery, it would probably be buried at the indicated site. In the illustrated example (fig. 5), plugs are cut out of the surface rock and replaced over the databank and the surface is roughened to conceal the location, so that even a close-up photograph from a space probe might not reveal anything unusual.

What would such a databank contain? Saunders believes that the main motive of an advanced civilization in leaving such an artifact would be humane. Its purpose would be to provide assistance to a civilization which is proceeding largely on good lines—peaceful, conservation-minded—but which might still destroy itself. In the illustrated example, the databank begins with decoding instructions and then continues with several layers of data. Some kind of interactive process might be incorporated so that the databank yields its treasures only to those most deserving.

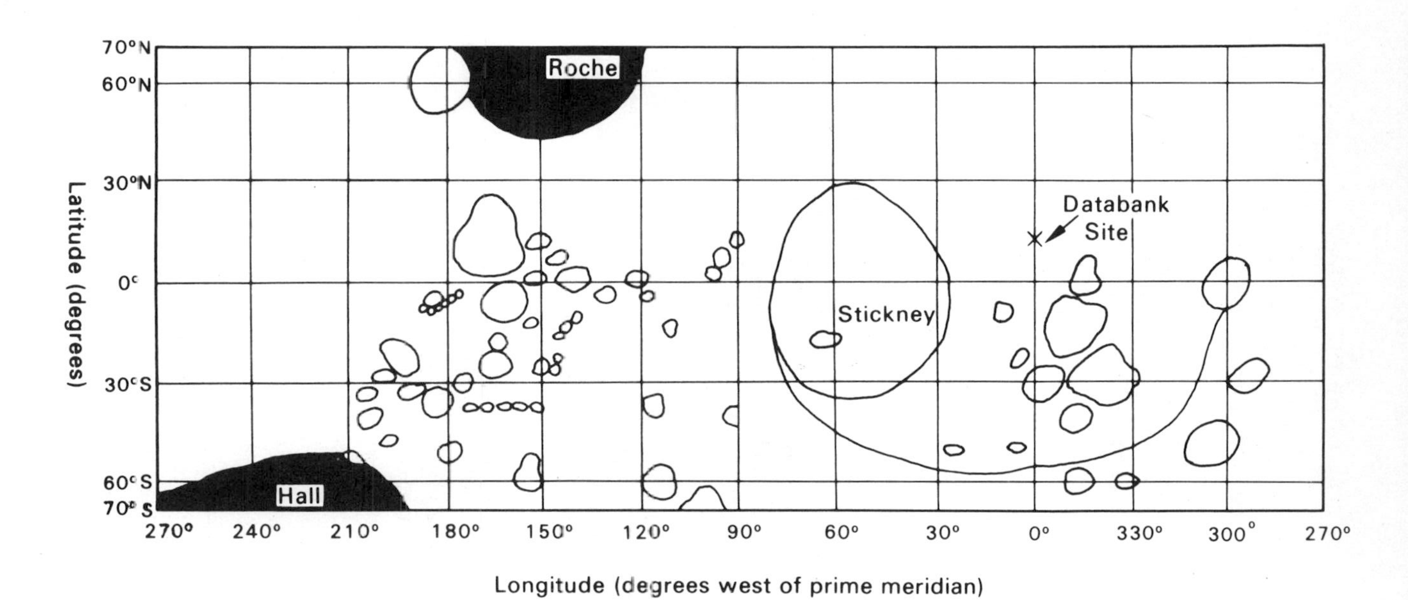

Figure 4. Location of databank on Phobos (Phobos map by Thomas C. Duxbury of the Jet Propulsion Laboratory).

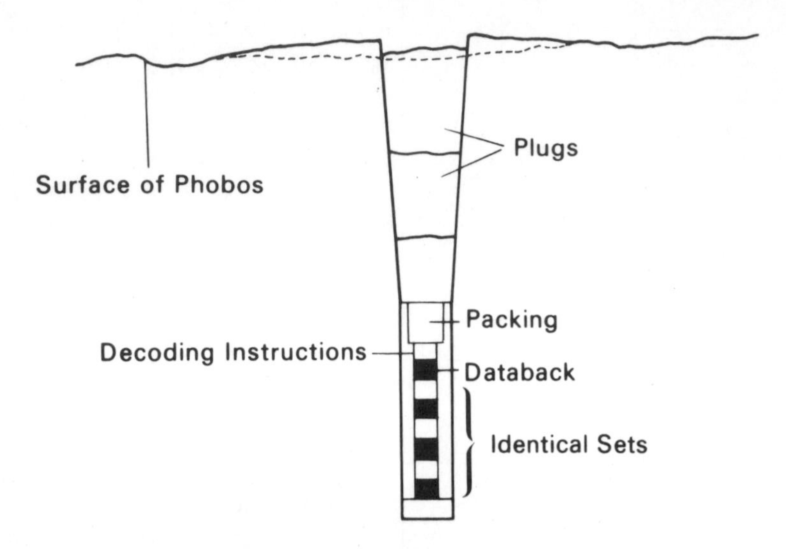

Figure 5. Possible construction of databank on Phobos.

Among the secrets to be imparted might be information on the likelihood of biological, nuclear, climatic, or other catastrophes, possibly drawn from the annals of galactic history. Instructions might also be included for receiving direct communications from the stars. In short, the hypothetical databank on Phobos would contain all the information of a radio message from the stars, or of a Bracewell messenger probe, but we would find it only after passing the message test left for us.

Saunders began to circulate his ideas about the Great Pyramid–Mars connection in a booklet published in 1975, and has since discussed the logic behind the use of such monument pointers in the *Journal of the British Interplanetary Society*. He agrees that it is by no means proven that Earth was visited by extraterrestrials, or that they deposited a databank on Phobos. So is the whole affair a result of what Carl Sagan terms the "statistical fallacy" of the enumeration of favorable circumstances—a polite name for coincidence hunting?

Saunders notes that Mars has only two moons, and the chain of coincidence involves both. The largest precision pyramid on Earth is involved, as also are the largest volcanoes on Earth and Mars. There are many more indicators to Mars and its moons than mentioned in this chapter; only the most fundamental have been highlighted. In an interview for this book, Saunders said that for every three possible coincidences he looked at, he found one that worked. The coincidences did not indicate any other planets or satellites or even Earth itself, and neither did the dimensions of the thirty or so other pyramids in Egypt. One of the weirdest coincidences of the whole affair is that Cairo, the site of the pyramids, was originally named El-Kahira, from the Arabic El-Kahir, meaning Mars.

Although the coincidences are 99 percent accurate or better, it must be remembered that the relevant distances and dimensions can be measured to far greater precision than this. For instance, the Great Pyramid's base length is 230.364 meters; 1 billion times this is 230.364 million kilometers. Yet the average Sun–Mars distance is 227.94 million kilometers, which is nearly 2.5 million kilometers smaller. The agreement may be 99 percent, but the actual error is immense. It seems surprising that advanced extraterrestrials did not design the pyramid so that the indicators were precise. The wider the errors that are allowed, the greater the possibility that the agreements are purely coincidental.

Scientist and writer Martin Gardner in his book *Fads and Fallacies in the Name of Science* has criticized the number-juggling exercises of various Pyramidologists, extending back to the last century:

> If you set about measuring a complicated structure like the Pyramid, you will quickly have on hand a great abundance of lengths to play with. If you have sufficient patience to juggle them about in various ways, you are certain to come out with many figures which coincide with important historical dates or figures in the sciences. Since you are bound by no rules, it would be odd indeed if this search for Pyramid 'truths' failed to meet with considerable success. The ability of the mind to fool itself by an unconscious 'fudging' on the facts—an unconscious overemphasis here and underemphasis there—is far greater than most people realize. The literature of Pyramidology stands as a permanent and pathetic tribute to that ability.

Robert Forrest, an English mathematician, has investigated the numerical relationships in other pyramids to put the coincidences concerning the Great Pyramid into perspective. He concentrated on Chephren's Pyramid, the second largest pyramid at Giza, which has a slope of 4/3 (53.13°). The ratio of the orbital periods of Saturn's satellites Hyperion and Titan is also 4/3, as indeed is the ratio of the orbital sizes of two satellites of Uranus, Oberon and Titania. To back this up, Forrest finds that 10 billion times the height of Chephren's Pyramid gives the mean distance of Saturn from the Sun, and the same factor times twice the pyramid's height gives the mean distance of Uranus. There are more links with Saturn, its satellites, Uranus, and other planets and their moons to be found in the dimensions of Chephren's Pyramid, although it must be said that none of these add up to such a consistent case as the relationships between the Great Pyramid and Mars. Nonetheless, as Forrest notes, these results show that it is relatively easy to invest the dimensions and proportions of any pyramid with unintended planetary associations.

Forrest also asks why, among all its useful data, the Great Pyramid tells us lots of apparently useless things as well. For instance, there are many ratios concerning the displacement of the Main Chamber and the access corridor in the pyramid that do not have a place in the Pyramid-Mars theory. If the Great Pyramid was designed as a pointer, Forrest argues, it is a poor one. It is too ambiguous and there are too many ways it can be misused. For instance, there are four faces and four edges to the pyramid, but the Saunders interpretation uses only one. Why are the others there, if all they can do is needlessly obscure the significant sight line?

If we use the two edges of the north face as sight lines extending into space, they cut the Earth's equatorial plane at a height of about 24,000 kilometers. This is twice the diameter of Venus, or half that of Neptune. But if the four edges of the pyramid were extended downward, Forrest calculated where they would emerge on the far side of Earth. The four points of emergence would, he says, make excellent hiding places for secret caches, as two are at the bottom of the Atlantic Ocean, one at the bottom of the Indian Ocean, and the fourth in the Gobi Desert. A less ambiguous monument might be a wedge-shaped structure, in which there was only one plausible sight line.

In short, thinks Forrest, the planetary associations of the pyramids prove nothing about the activities of extraterrestrials, but they do say something about the way human imaginations can make pyramids fit theories. The Great Pyramid, he says, has "proved" too many theories for its own good. Like the numerous alternative authors suggested for

Shakespeare's plays, once the same data has been used to "prove" the case of several claimants, one begins to realize the limitations of each.

It is impossible to prove or disprove Saunders' theory, because the Pyramid–Mars coincidences really do exist; they cannot be dismissed like the inventions of von Daniken and his ilk. Whether or not the coincidences are convincing is subjective. The only conclusive test would be an actual expedition to Phobos, and that is unlikely to take place in the near future unless space-probe photographs show something untoward at the indicated site.

In the meantime, there stands the Great Pyramid, a monument not, perhaps, to an extraterrestrial visitation, but rather a monument to the laws of chance.

REFERENCES

Saunders, M. W., *Destiny Mars* (Caterham, Surrey: Downs Books, 1975).
——— *Journal of the British Interplanetary Society*, Vol. 30 (1977), p. 349.

13

The Amphibians of Sirius

Amphibious beings from the star Sirius, 8.7 light-years away, visited the Earth 5,000 or more years ago, leaving advanced astronomical knowledge that is still possessed by a remote African tribe. That is the startling essence of perhaps the most puzzling of all ancient-astronaut stories, first popularized by writer and Orientalist Robert Temple in his 1976 book *The Sirius Mystery* and since adopted by Erich von Daniken. At the center of the mystery are the Dogon people living near Bandiagara, about 300 kilometers south of Timbuktu in Mali, western Africa, knowledge of whose customs and beliefs we owe to the French anthropologists Marcel Griaule and Germaine Dieterlen, who worked among the Dogon from 1931 to 1952.

Between 1946 and 1950 the Dogon head tribesmen unfolded to Griaule and Dieterlen the innermost secrets of their knowledge of astronomy, which is shared by the neighboring peoples the Bambara and the Bozo. Much of this secret lore is complex and obscure, as befits ancient legends, but there are certain specific facts which stand out. For instance, the Dogon know that Jupiter has four moons, that Saturn has rings, and that there are endless stars in the sky. They know that Earth spins on its axis and that it, along with the other planets, orbits the Sun. Motions of objects in the sky are likened by the Dogon to the circulation of the blood, about which they also seem to have remarkably detailed knowledge—including references to the processes of respiration and digestion and of red and white blood cells.

But their most remarkable knowledge concerns the star Sirius, with which their religion and culture are deeply concerned. In their information imparted to the French anthropologists, the Dogon refer to a small, superdense companion of the star Sirius, made of matter heavier than anything on Earth, and moving in a 50-year elliptical orbit around its parent star. Sirius does indeed have a companion star answering this description: a white dwarf, invisible without a telescope. The white-dwarf companion of Sirius was not seen until 1862 when the American optician Alvan Graham Clark spotted it while testing a new telescope. But the Dogon tradition concerning a companion of Sirius indubitably extends back thousands of years, well before any terrestrial astronomer could have known of its existence, let alone its composition and orbital details. How can we account for the remarkable accord between the ancient Dogon legends and modern astronomical fact?

A Dogon legend, similar to many other tales by primitive people of visits from the sky, speaks of an "ark" descending to the ground amid a great wind. Robert Temple interprets this as the landing of a rocket-powered spacecraft bringing beings from the star Sirius. According to Dogon legend, the descent of the ark brought to Earth an amphibious being, or group of beings, known as the Nommo. "Nommo is the collective name for the great culture-hero and founder of civilization who came from the Sirius system to set up society on the Earth," explains Temple. Why should the Nommo be amphibians? He notes that amphibious beings would have an advantage on a planet around a hot star like Sirius, because the water would keep them cool and also because water absorbs the dangerous short-wavelength radiation that hot stars like Sirius emit more strongly than the Sun.

Temple equates the mythical Nommo with the fish-god Oannes whom the Babylonians credited with the founding of their civilization. This similarity between Dogon traditions and those of Middle Eastern or Mediterranean civilizations is no surprise, as the Dogon share common cultural and physical roots with those peoples. The ancient Egyptians were also preoccupied with Sirius; they even based their calendar on its yearly motion around the sky. That fact is easily explained without recourse to ancient astronauts, because Sirius is the brightest star in the entire sky, easily visible from Mediterranean regions. The first observable rising of Sirius before the Sun (termed the heliacal rising) marked the beginning and end of each Egyptian year; it also coincided with the agriculturally all-important flooding of the Nile. No wonder that the Egyptians accorded Sirius such respect, regarded it as revitalizing life on Earth, and embodied it in their legends. Sirius would be assured of this honored place in Egyptian and Dogon belief irrespective of Temple's ancient-astronaut theory. So is there any independent evidence of life around Sirius to back up Temple's interpretation of the Dogon legends?

First, let's look at Sirius and its companion star, based on what we have already said about stars and planets in this book, to see if it is at least theoretically plausible that advanced life might have arisen in the Sirius system. Sirius A, the bright white star we see from Earth, has a mass 2.35 times that of the Sun. Its white-dwarf companion, Sirius B, has a mass 0.99 that of the Sun. Stellar evolutionary theory tells us that the most massive stars burn out the quickest, so originally Sirius B must have been the more massive star, before burning out to become a white dwarf. Probably Sirius B spilled over some of its gas onto Sirius A during its aging process, so that the original masses of the two stars were approximately the reverse of what we see today.

Let us assume that Sirius B initially had a mass at least twice that of the Sun (it could have been more), while Sirius A had about one solar mass (the sum of their masses must have been at least 3.34 Suns, which is their total mass today). A star with twice the Sun's mass lives for no more than about 1 billion years before swelling up into a red giant, as the Sun will do at the end of its life; this does not seem long enough for advanced life to develop. Even if life had evolved, the red-giant stage of Sirius B would have led to its disappearance. As the star swelled up there would be changes in climate on any nearby planet, followed by a stellar gale between the two stars as hot gas streamed from Sirius B onto Sirius A. Theorists calculate that during this mass transfer the two stars would have moved apart; so they were originally much closer than we see them today.

Astronomer Stephen Maran of NASA's Goddard Space Flight Center has reviewed information on Sirius, and finds from spectroscopc observations that the surface layers of Sirius A do indeed seem to have been contaminated with material from Sirius B. Perhaps Sirius B lost more than half its original mass. Some of its gas would have been lost forever into space, but much of it would have been claimed by Sirius A, roughly doubling that star's mass and leaving the core of Sirius B as the burned-out white dwarf we see today.

There is an interesting additional problem here. Astronomers of Ptolemy's time, 2,000 years ago, referred to Sirius as red, while we know it to be almost pure white. Is it possible that they were seeing the end of Sirius B's red-giant phase? And is it possible even that the Oannes/Nommo creatures came to Earth many thousands of years ago to escape the impending doom of their home star? Alas, this charming speculation is not supported by facts. For one thing, theorists calculate that the mass-transfer phase between the two stars would have lasted at least 100,000 years, and could have taken millions of years. The most reliable observations of the size and temperature of Sirius B, analyzed by theorist H. L. Shipman of the University of Delaware, show that it must have been a cooling-down white dwarf for at least 30 million years. Probably the Greeks described Sirius as red because they observed it close to the horizon, where atmospheric effects make it glint red.

If we are to take the Dogon legend at face value, it seems that the Nommo visited Earth after Sirius B became a white dwarf. But it is even less likely that life could have arisen in the Sirius system since that time. Sirius A can only live for a few hundred million years in its present form before it too burns out (and its own red-giant death will be a spectacular thing to see from Earth). There has not been time for life to arise in the Sirius system since Sirius B became a white dwarf. In any case, life there now would not be too healthy, for Dutch X-ray astronomers have found that the Sirius system emits soft X rays, which H. L. Shipman has demonstrated must originate from deep in the layers of hot gas around Sirius B. And to cap it all, Robert S. Harrington of the U.S. Naval Observatory has shown that habitable planets cannot exist in stable orbits in the Sirius system.

In short, astronomical evidence argues strongly against Temple's ancient-astronaut theory.

With this information in mind, let's now look in more detail at the Dogon legend. Immediately, we encounter a surprise: Dogon legend maintains that Sirius has *two* companions. This belief comes not from astronomy but from an important concept of twinning which the

Dogon and their related peoples apply to many objects, not just to Sirius. According to the twinning concept, two companions representing the male and female sexes accompany Sirius like the two horns on either side of the head of an animal. On the BBC-TV Horizon documentary on ancient astronauts Mme. Dieterlen showed centuries-old Dogon ritual masks to illustrate this concept. The masks consist of a head with horns, the head representing Sirius and the horns the two companions. Also on the masks are symbols representing Sirius and its two companions, one on either side.

The companion identified as Sirius B is called *po tolo,* which Griaule and Dieterlen term the Digitaria star (*tolo* means star, and *po* is the Dogon name for a cereal seed, *Digitaria exilis*). The second star is *emme ya,* sorghum-female, which according to Griaule and Dieterlen is the seat of the female souls of all living or future beings. Digitaria assumes the greater importance of the two Sirius companions because it has a symbolic association with the male circumcision ceremony. As in so many cultures, the complex Dogon mythology is largely concerned with male-female sexuality (witness the attributes of Digitaria and *emme ya*) and of fertility. Sirius has been associated with creation and revitalization since the time of the ancient Egyptians, so these attributes are scarcely mysterious.

Is there any astronomical evidence for a third star in the Sirius system? Several astronomers in the 1920s and 1930s suggested the existence of a possible third star because of an apparent wobble in the measured orbits of Sirius A and B. Astronomer Philip Fox believed he had seen a close companion of Sirius B in 1921 at Dearborn Observatory, Illinois and further sightings of this supposed Sirius C came a few years later from observers in South Africa. No one else managed to see it, though. In 1965 Irving W. Lindenblad at the U.S. Naval Observatory, Washington, D.C., began a series of photographic observations of Sirius A and B to solve the problem. Lindenblad reported in 1973 that his observations, more accurate than any before, showed no evidence of a close companion to either Sirius A or B.*

Since this point is so important, I wrote to Dr. Lindenblad for addi-

* Temple wrongly claimed that the first photograph of Sirius B was taken by Lindenblad in 1970. In fact, photographs of Sirius B were taken by G. B. van Albada at the Bosscha Observatory in Indonesia from 1955 onward, and also at Sproul Observatory, Swarthmore, Pennsylvania, from 1964.

tional confirmation. He replied: "My work disproved ideas that had persisted for years, namely that analysis of the visual observations made since the discovery of Sirius B gave evidence of a perturbation caused by some third body. The possibility of a *very distant* third body cannot be ruled out theoretically as being physically impossible but there is absolutely no evidence for such a body."

What do the Dogon say about their imaginary third member of the Sirius system, which they call *emme ya?* There are two sources of information. One is an article entitled "A Sudanese Sirius System," published by Griaule and Dieterlen in 1950. The second is a book by the two called *Le Renard pâle*, published in 1965, which gives a more detailed description of Dogon astronomy. According to Griaule and Dieterlen, the Dogon describe *emme ya* as larger than Digitaria, but four times as light in weight, a description which is consistent with a red dwarf. Yet such a body anywhere near Sirius A and B would have shown up in Lindenblad's observations. In the "Sudanese Sirius System" article, Griaule and Dieterlen report that *emme ya* supposedly travels along a greater trajectory than Digitaria, but in the same time. This is a physical impossibility. According to Kepler's laws, the larger an orbit, the longer an object takes to go around it. So here the legend is astronomically wrong.

Contradictorily, in *Le Renard pâle* it is said that the orbital period of *emme ya* is 32 years, shorter than the 50-year period of Digitaria, which would mean that *emme ya*'s orbit is smaller than that of Digitaria, not greater. Temple has recently suggested that the figure of 32 years actually refers to close approaches between *emme ya* and Digitaria, which would occur every 32 years if *emme ya* moved along a larger orbit every 100 years in the opposite direction to Digitaria in its 50-year orbit. But this introduces another conflict, because the legend specifically says that the two bodies travel in the same direction. The respective positions of Digitaria and *emme ya* are said in one part of Griaule and Dieterlen's writings to be at right angles, and elsewhere as in a line. Where the legend is not at variance with fact, it is self-contradictory. On this evidence, the Sirius mystery is intractable.

If the Dogon knowledge of Sirius C is unreliable, what do they say about Sirius B (Digitaria)? It is described as being the smallest and heaviest star, consisting of a heavy metal known as *sagala*. It was certainly the smallest and heaviest star known in the 1920s, when the superdense nature of white dwarfs was becoming understood; the material of which white dwarfs are made is indeed compressed more densely than metal. But since then many other white dwarfs have been found, not to mention neutron stars, which are far smaller and denser. Any visiting spaceman would certainly have known about these, as well as black holes; he would also have known that Jupiter has more than four moons.

The Dogon are supposed to know that Sirius B orbits every 50 years. But what do they actually say? Griaule and Dieterlen record it as follows: "The period of the orbit is counted double, that is, one hundred years, because the Siguis are convened in pairs of 'twins,' so as to insist on the basic principle of twinness." Temple emphasizes that the number 50 appears regularly in ancient myths, such as the 50 companions of Gilgamesh, but it is difficult to see what this has to do with the Dogon. The Sigui ceremony referred to above is a ceremony of the rebirth of the world that is celebrated every 60 years (originally it was celebrated every seven years). Griaule and Dieterlen admit that the 60-year recurrence of the Sigui ceremony is inconsistent with the 50-year orbit of Sirius B.

The Dogon are also supposed to know that Sirius B orbits Sirius A in an ellipse. Where does this information come from? Actually, it comes from Robert Temple. At a yearly ceremony known as the *bado*, the Dogon make a sand drawing of the Sirius system (fig. 6). This time they place *emme ya*, "the sun of women," at the center. Around it are marked Sirius, represented by a cross; Digitaria, shown in two positions, drawn as a horseshoe shape to indicate its nature as a collector and distributor of matter placed in it by the Creator; and five other signs representing different objects, one of them the Nommo. Drawn around these symbols is an oval, the egg of the world. The oval is a device which the Dogon use to enclose other diagrams, not just of Sirius.

The Dogon sand diagram of the complete Sirius system, as described by Griaule and Dieterlen, contains clear elements of male-female sexuality; it is symbolic. Temple chooses to interpret it literally. On pages 23 and 25 of his book he gives his own modified version of the diagram. Temple retains the symbol for Sirius, plus one of the positions of Digitaria, and the surrounding oval; he omits all the other symbols that fill the oval. He then interprets the "egg of the world"

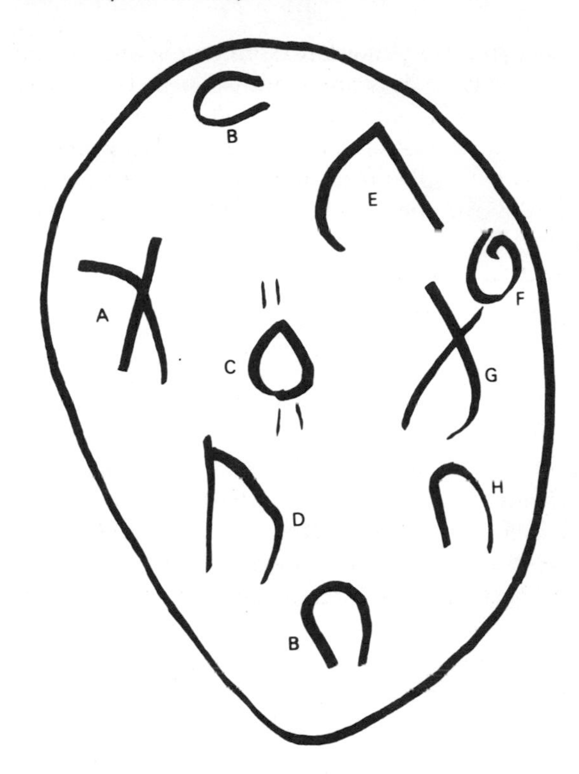

Figure 6. Dogon sand drawing of the complete Sirius system, after Marcel Griaule and Germaine Dieterlen. *A*, Sirius; *B*, Digitaria, shown in two positions; *C*, *emme ya*, the sun of women; *D*, the Nommo; *E*, the Yourougou, a mythical male figure whose destiny is to pursue his female twin; *F*, the star of women, a satellite of *emme ya*; *G*, the sign of women; *H*, the sex of women, represented by a womb shape. The whole system is enclosed in an oval, representing the egg of the world.

oval as the orbital path of Sirius B around Sirius A, even though Digitaria is drawn as lying within it, not on it. This, then, is the basis for saying that the Dogon "know" the orbit of Sirius B around Sirius A to be an ellipse.

Of course, the most critical factor of all for the ancient-astronaut theory is that the Dogon say the Nommo came from Sirius. Where do they say this? Again, Temple says it for them. On page 217 of his book, he reports that the Dogon say that "*po tolo* and Sirius were once where the Sun now is." This ambiguous statement is the only quotation he offers throughout the entire 290-page book to substantiate this most vital claim. He shows four designs, looking like Cleopatra's Needle, found on Dogon masks and speculates: "Maybe the Dogon have actually drawn a rocket ship." In fact, the pencil-shaped structures, over seven times as tall as they are broad, would be too unstable for a landing craft. Nowhere does the Dogon legend specifically state that the Nommo came from Sirius.

And what of the nature of the Nommo itself (or themselves)? The Nommo is a spiritual being, associated with the life-giving rains that are vital in this arid part of the world. An interesting sidelight here is that the neighboring Bozo people, who share much of the Sirius mythology, believe that the lungfish which lives in the mud of the Niger River "falls from heaven with the first rainstorm of the season." This sounds to me a lot like the Nommo, but not like an extraterrestrial visitation. The obsession with fishlike creatures is no surprise, since fishing on the Niger River is a major industry hereabouts; the town of Mopti on the Niger, about fifty kilometers from Bandiagara, is the area's commercial center.

Looked at in this light, Temple's ancient-astronaut theory for the Sirius mystery is no more solid than his assumptions and interpretations of the Dogon legend—a legend which, as we have seen, is riddled with contradictions, at least when one attempts to interpret it in terms of the real world.

In his book, Temple makes one prediction which allows a test of his theory. He asks: "What if this is proven by our detecting on our radio telescopes actual traces of local radio communications?" I asked two astronomers engaged in the search for extraterrestrial radio messages, Paul Feldman at the Algonquin Observatory in Ontario and Robert Dixon at Ohio State University to help (they would otherwise have paid no attention to Sirius because of its extreme unlikelihood of supporting life). Feldman's search was filmed by BBC-TV for the Horizon documentary on ancient astronauts. In April 1977 Feldman listened to Sirius with the Algonquin dish for three ten-minute periods; no signal was detected. Simultaneously with Feldman's search, but at a different wavelength, Robert Dixon in Ohio listened to Sirius for one-minute intervals on each of thirteen days in April 1977; again no signals were detected. So two independent tests of Temple's hypothesis have drawn a blank. It is difficult to resist the conclusion that there is no life in the Sirius system, nor has there ever been.

The true test of a good extraterrestrial story is that it should tell us something we don't already know. The Dogon legend only tells us what we know already; and some of what it does tell us is wrong.

Nevertheless, there are certain indisputable facts, such as concerning the rings of Saturn, the moons of Jupiter, and the circulation of the blood, that the Dogon could not plausibly have discovered for themselves. Who told them?

It is all too easy for Westerners to think of African tribes as isolated, ignorant, and uneducated. If the Dogon have their roots in the Greek and Egyptian civilizations, as seems to be the case, then they are certainly not ignorant. Neither are they isolated or uneducated. The Dogon live near an overland trade route, as well as close to the southern banks of the Niger River, which is another channel of trade. Any number of travelers could have passed through their midst, or Dogon tribesmen could themselves have journeyed far, possibly meeting astronomically informed seamen on the coast. The Dogon have been in contact with Europeans since at least the late nineteenth century. It is even said that members of the Dogon fought for the French in the trenches during the First World War.

Astronomers Peter and Roland Pesch of the Warner and Swasey Observatory in Ohio have pointed out that French schools have existed in the Dogon area since 1907, including geography and natural history in their curricula. Dogon members wishing to pursue higher education have been able to do so in nearby towns. There are also Islamic schools, and the Dogon have apparently incorporated aspects of Islamic ritual and culture into their own culture.

Then there are missionaries, who would naturally be closely interested in the legends of the natives. I confirmed with the London headquarters of the White Fathers, a Catholic group who have been very active in this part of Africa, that missionaries from their sect had made contact with the Dogon in the 1920s. It is tempting to speculate that certain of the more specific details about Sirius B were grafted onto the existing Sirius legend at that time, because it was about then that astronomers had just discovered the true nature of Sirius B as a tiny, superdense star, and white dwarfs were being accorded the same kind of publicity as attends black holes today. Alas, in the missionaries' summary reports of their activities there is no mention that they discussed Sirius with the Dogon, although if more detailed notes were published they might add significantly to our understanding of the origins of Dogon knowledge.

The point is that there are any number of channels by which the Dogon could have received Western knowledge long before they were visited by Griaule and Dieterlen. Dogon knowledge of the moons of Jupiter, the rings of Saturn, and the circulation of the blood confirms that some such influx of information has taken place. Although the Sirius mythology of the Dogon, particularly the story of Nommo and the concept of twinning, is admittedly both ancient and profound, is it not at least possible that some of the more superficial resemblances to astronomical fact are trimmings added in this century?

"This full cycle return of a myth to its culture of origin through an unwary anthropologist might sound unlikely if there were not so many examples of it in anthropological lore," says Carl Sagan. He recounts the true story of an anthropologist who was investigating the traditions of American Indians early in this century. The anthropologist asked one of the local elders about rituals and ceremonies concerning childbirth, puberty, marriage, and death. Each time, the elderly informant retreated into his hogan before emerging with the answer. Could he, the anthropologist wondered, be consulting an even more aged Indian who lived within? In fact, it turned out that he was looking up the answers in the *Dictionary of American Ethnography*.

Perhaps of more relevance to the Sirius case are two examples concerning the physician Carleton Gajdusek, who in 1976 shared the Nobel Prize for Medicine. In 1957 Dr. Gajdusek and some companions traveled into New Guinea, where they stayed briefly with some of the native inhabitants. One night the native hosts sang some of their traditional songs, and in return Gajdusek's party sang some traditional Russian songs, including "Ochichornia," of which the natives requested many repetitions. Several years later, while engaged in collecting traditional songs in a nearby region of New Guinea, Gajdusek was amazed to hear a slightly altered but still recognizable version of "Ochichornia," which had by then become accepted as a traditional native song.

Shortly thereafter, Gajdusek was visited by an Australian physician who had found that some New Guinea natives believed that a certain disease was transmitted in the form of an invisible spirit that entered the skin of a patient. With a stick, the native informant had sketched in the sand a circle, outside which, he explained, was black, and inside which was light. Within the circle he drew a squiggly line to represent the appearance of these invisible malevolent spirits. How did the natives get such an astounding insight into the transmission of disease by microbes? Years earlier, Gajdusek had shown the natives a disease-causing germ under his microscope, and the sand drawing was simply the natives' recollection of this deeply impressive sight.

"All three of these stories underline the almost inevitable problems encountered in trying to extract from a 'primitive' people their ancient legends," notes Sagan. "Can you be sure others have not come before you and destroyed the pristine state of the native myth?"

In view of the Dogon fixation with Sirius it would surely be more surprising if they had *not* grafted onto their existing legend some new astronomical information gained from Europeans, picking what fitted their purpose and ignoring the rest. We may never be able to reconstruct the exact route by which the Dogon received their current knowledge, but out of the confusion one thing at least is clear: they were not told by beings from the star Sirius.

REFERENCES

Harrington, R. S., *Astronomical Journal*, Vol. 82 (1977), p. 753.
Lindenblad, I. W., *Astronomical Journal*, Vol. 78 (1973), p. 205.
Maran, S. P., *Natural History*, August–September 1975, p. 82.
Pesch, P., and R. Pesch, *The Observatory*, Vol. 97 (1977), p. 26.
Shipman, H. L., *Astrophysical Journal*, Vol. 206 (1976), p. L67.
Temple, R. K. G., *The Sirius Mystery* (London: Sidgwick and Jackson, 1976).

14

The Siberian Spaceship Explosion

Siberia, June 30, 1908. On that summer's morning, a fireball, described as "too bright for the naked eye," descended to Earth in the valley of the Podkamennaya Tunguska [Stony Tunguska] River (60°55′N, 101°57′E), 800 kilometers northwest of Lake Baikal. At the end of its path it disintegrated explosively with a force rivaled only by the greatest natural disasters, such as the eruption of the volcano Krakatoa in 1883. Its effects were seen and heard over an area about 1,500 kilometers diameter.

The great Siberian fireball of 1908 was an event so exceptional that it excited a controversy that continues to this day. Explanations offered for it go outside the normal range of astronomical phenomena into the realm of the bizarre, including the remarkable hypothesis that it was caused by nothing less than the crash-landing of a nuclear-powered spaceship. We shall examine the case in detail here, for it is often put forward as being among the best evidence for a visit by extraterrestrials in modern times.

The area on which the object fell was sparsely inhabited by the Tungus, a hardy, Mongol-like people who herded reindeer. No one was killed by the blast, but the catastrophic nature of the object which caused it is evident from eyewitness reports.

A farmer on his porch 60 kilometers away described how he saw toward the northwest a fireball that "covered an enormous part of the sky." The heat from it seemed to be burning his shirt, and a neighbor clasped his hands over his ears, which felt as though they were burning. The blinding bright-blue bolide, trailing a column of dust, exploded with a force that knocked the first farmer off his porch, where he lay unconscious for a few seconds. On coming to, he felt ground tremors that shook the entire house, broke the barn door, and shattered windows. In the house of his neighbor, earth fell from the ceiling and a door flew off the stove. Sounds like thunder rumbled in the air.

Farther north, nearer the center of the fall, several of the nomadic Tungus were thrown into the air by the explosion, and their tents were carried away in a violent wind. Around them, the forest began to blaze.

As the dazed Tungus cautiously inspected the site of the blast they found scenes of terrifying devastation. Trees were felled like matchsticks for up to thirty kilometers around. The intense heat from the explosion had melted metal objects, destroyed storehouses, and burned reindeer to death. No living animals were left in the area.

Other effects were noted around the world, but their cause long remained a puzzle because news of the fireball and explosion did not become widely known for many years. Seismic waves like those from an earthquake were recorded throughout Europe, as well as disturbances of the Earth's magnetic field. Meteorologists later found from microbarograph records that atmospheric shock waves from the blast had circled the Earth twice.

Eerie nighttime effects were noted in western Asia and Europe immediately after the Tunguska fall. Reports over this area speak of nights up to a hundred times brighter than normal, by which it was possible to read at midnight, and crimson hues in the sky, particularly toward the north, like the glare from fires. The strange lights did not flicker or form arches like aurorae, which are caused by ionized air; instead, they were compared with effects that followed the outburst of Krakatoa, which injected vast clouds of dust into the atmosphere.

In the United States, the astronomer Charles Greeley Abbot, noted for his long-term monitoring of the Sun's energy received on Earth, found that the transparency of the atmosphere was noticeably lowered from the middle of July 1908 until late August; this is now attributable to vast amounts of dust from the Tunguska object that had been distributed by upper-atmosphere winds.

At the time of the Tunguska fall, Russia was entering a period of major political upheaval. Despite the exceptional nature of the Tunguska event, news about it remained buried in local Siberian newspapers until accidentally uncovered thirteen years later by a Soviet mineralogist, Leonid Kulik.

In the hope of finding a massive and hitherto unknown meteorite, possibly made of iron, Kulik traveled to Kansk on the Trans-Siberian Railroad in September 1921. When he arrived he learned that the site of the fall had been several hundred kilometers to the north, across frozen and desolate taiga. Unable to proceed farther, he collected eyewitness reports which convinced him that a massive meteorite had indeed passed overhead, moving from south to north, toward the area of the Stony Tunguska River.

These reports, and those from other scientists who spoke to the inhabitants of the region, seemed at last to explain the hitherto baffling records of seismic and acoustic disturbances on June 30, 1908. Kulik set out on a major expedition in 1927, quite naturally expecting to find the site of a meteorite impact that would rival the one-kilometer-wide Arizona crater formed in prehistoric times.

Instead, Kulik uncovered the great puzzle of the Tunguska explosion: despite the immense devastation caused by the blast wave, there was no evidence that the object itself had reached the ground. Instead of a massive crater, he found a frozen swamp and a curious stand of trees which, despite being at the center of the explosion, had escaped the effects of the blast that had leveled all around them. Trees and ground had been scorched by the intense heat of the explosion out to a distance of eighteen kilometers.

Kulik returned again in 1928 and 1929–30 in search of meteorite fragments, but found nothing. His fruitless searches showed that the object could not have been made of iron, as he had originally supposed. So what was the object that caused such destruction by the banks of the Stony Tunguska?

In 1930 the English meteorologist Francis J. W. Whipple, assistant director of the Meteorological Office, proposed that the Tunguska event had been caused by the collision of the Earth with a small comet, a suggestion supported by the Soviet astronomer A. S. Astapovich and backed up by many investigators since then.

The popular view of a comet is a giant glowing ball of dust and gas trailing streamers for millions of miles, as with the spectacular appearance of Halley's Comet in 1910. But such brilliant comets are the exception rather than the rule. As many as fifteen comets are tracked each year by astronomers, but none of them may become visible to the naked eye. Most comets are smaller and fainter than the spectacular objects illustrated in astronomy books; some comets, particularly old ones, show scarcely any tail at all.

In the vision of the American astronomer Fred L. Whipple (no relation to the English meteorologist), a comet resembles a dirty snowball of frozen gas and dust. The evaporating gases keep the comet active by releasing frozen-in dust particles and producing the long, flowing tail. As the comet runs out of gas it declines to become nothing more than a loose bag of low-density rocks. Such an object would indeed cause a blazing fireball as it began to burn up by friction in passing through the Earth's atmosphere, eventually disintegrating explosively as the deceleration forces overcame its own internal strength.*

As with any puzzling event, there have been a host of alternative explanations. As recently as 1973, physicists A. A. Jackson and Michael P. Ryan of the University of Texas suggested that a mini–black hole had blasted into Siberia. According to some astronomers, such mini–black holes, with the mass of an asteroid but the size of an atomic particle, might have formed in the conditions of intense pressure and high density shortly after the Big Bang explosion which marked the beginning of the Universe as we know it. The passage of a mini–black hole through the atmosphere would, Jackson and Ryan calculated, have all the observed effects of the Tunguska fireball. However, the mini–black hole would have carried right on through the Earth and emerged in the North Atlantic, producing similar spectacular effects as it departed. No such effects were recorded.

But the most controversial theory about the Tunguska event was proposed in 1946 by the Soviet author Alexander Kazantsev, in whose

* Note that the energy of the event is supplied purely by the object's own high-speed motion, not by any chemical or nuclear combustion. For the technically minded, the physics of meteorite explosions, with particular reference to Tunguska, have been modeled by the Soviet specialist V. P. Korobeinikov et al. (*Astronautica Acta*, Vol. 17 [1972], p. 339, *Acta Astronautica*, Vol. 3 [1976], p. 615) and L. I. Turchak (*Soviet Physics Doklady*, Vol. 21 [1976], p. 351).

mind the atomic-bomb devastation of Hiroshima remained fresh. The apparent similarity between the blast effects at Hiroshima and at Tunguska led Kazantsev to suppose that the explosion over Siberia had been caused by the burn-up in the atmosphere of a nuclear-powered spaceship from Mars. Presumably, thought Kazantsev, the crew had come to pick up water for their parched planet from Lake Baikal, the largest volume of fresh water on Earth, but had perished during the journey. As the craft plummeted uncontrollably into the Earth's atmosphere it heated up by friction until the power plant erupted in a midair blast like the Hiroshima bomb.

The idea was put forward as a science fantasy story. But it gained lasting popularity—and even a superficial measure of scientific support.

For one thing, the midair blast would explain why there was no crater or meteorite fragments, and it would also account for the characteristic clump of trees that remained standing at the center of the Tunguska devastation, as they had under the explosion point of the Hiroshima bomb. The tremendous heat output of the Tunguska explosion, the force of the shock wave, and the updraft of heated air that caused a "fiery pillar" have all been compared with the effects of the Hiroshima blast.

Aerodynamicist Felix Zigel of the Moscow Institute of Aviation, a supporter of the nuclear-powered-spaceship theory, even suggested that the craft had performed a maneuver in the atmosphere before it crashed. He based his contention on analysis of eyewitness reports of the fireball, which seemed to indicate that the descending object initially approached from the south, whereas the blast pattern at the site, as evidenced by the uprooted trees, indicated a final approach from the east-southeast. Zigel proposed that the Tunguska body described a crazy right-left zigzag as it came in to land, first veering east and then doubling back to the west, even though no witnesses actually reported having seen it change course.

Zigel's course-change proposition depends on the accuracy of the trajectory reconstructed from eyewitness reports of the fireball as it descended over Siberia. These reports, of course, came from unskilled farmers and peasants, and were collected twenty years after the event. Howard Miles, a past president of the British Astronomical Association and a man experienced in the analysis of satellite and fireball reports, notes that unless such an object passes directly overhead, it is virtually impossible to assess its flight path accurately. Most of the eyewitnesses lived southwest of the blast site; the Soviet astrophysicist Vasilii Fesenkov, analyzing eyewitness reports, estimated that the fireball was from 20° to 40° above the horizon as seen from the towns of Kamenka and Kezhma southwest of the impact point. Therefore the evidence seems too unreliable to support Zigel's far-reaching conclusion. In any case, an approach from the southeast is not what one would expect for a spacecraft coming in to land. Rather, it would approach from the west, following the direction of the planet's rotation, instead of heading into it.

Another scientist who appeared to deny a natural origin of the Tunguska blast is former oilman Alexei Zolotov, now a geophysicist with the Soviet Academy of Sciences. In a scientific paper published in 1967 he maintained that the ballistic shock wave caused by the object's motion through the atmosphere was negligible, and that the blast wave which felled the trees at Tunguska was entirely due to the end-point explosion. On this basis, he concluded that the Tunguska object exploded "because of the internal energy of the meteorite itself." In other words, he did not believe it was a true "meteorite" at all—he actually favored the spaceship hypothesis.

But he ignored the results of an experiment performed the previous year by Igor Zotkin and Mikhail Tsikulin of the Soviet Academy of Sciences' meteorite committee, who reproduced the effects of the Tunguska blast by means of a scale model. They ignited an exploding cord, tilted at an angle above a field of model trees, to simulate the sonic-boom shock wave as the speeding object passed through the atmosphere, and with an additional explosive charge at the end point to simulate its final disintegration in the air. The terminal explosion left a knot of standing trees directly underneath, as observed at the Tunguska site.

Zotkin and Tsikulin found that the closest agreement between the simulated field of felled trees and the actual throw-down pattern observed at Tunguska occurred with the cord at an inclination of 30° to the horizontal and a fourfold amplification at the end point. They concluded that the fall pattern of the trees at Tunguska, which has characteristic butterfly-like "wings" that cannot be explained by a single central explosion alone, resulted from the combined effects of a ballistic wave caused by the object's supersonic motion through the atmosphere and a terminal explosion as it disintegrated. According to them, the object approached from the east-southeast (65° east of south).

Ari Ben-Menahem, a geophysicist at the Weizmann Institute of Science in Rehovot, Israel, reached similar conclusions in 1975 from a comparison of old seismograms of the Tunguska event with those of air explosions at the nuclear test sites at Novaya Zemlya and Lop-Nor. Ben-Menahem's analysis, probably the most accurate ever performed for the Tunguska event, established the local time of the blast as 7h 14m 28s, at an altitude of 8.5 kilometers. He agreed with the 65°-east-of-south approach path and the combination of ballistic and terminal-explosion shock waves, but favored a shallower angle of descent—about 15°. Without speculating on the nature of the Tunguska object, Ben-Menahem compared its effects with that of a nuclear missile with a 12.5-megaton warhead—an energy release equivalent to 5×10^{23} ergs.

There can be no doubt that the evidence indicates an aerial blast of enormous power. But was it a nuclear explosion? Some writers have followed Kazantsev in comparing the effects of the Tunguska explosion with those of the Hiroshima bomb—the strong thermal flash, the updraft of heated air that produced a mushroom-shaped cloud of black smoke, and the standing trees under the explosion point. But these effects are not unique to a nuclear blast. Any explosion is followed by an updraft of heated air and a puff of smoke. The thermal effects could just as easily have been caused by the frictional heating and ultimate explosive break-up of a large meteoric body plunging into the atmosphere. Brilliant exploding fireballs, similar to the effects of a satellite's re-entry into the atmosphere, are regularly caused as chunks of rock and metal from space burn up in the air; most of the solar system's debris is too fragile to reach the Earth's surface intact. And the clump of standing trees under the center of the blast, like that at Hiroshima, would have been left by an aerial explosion of any kind, as the scale-model experiments showed.

So are there any real grounds for the belief in a nuclear explosion over Siberia in 1908? Scientists have looked . . . but found nothing.

In a 1965 paper published in *Nature*, three American scientists, Clyde Cowan, C. R. Atluri, and Willard Libby, pointed out that a nuclear explosion releases a burst of neutrons, which turn atmospheric nitrogen into radioactive carbon-14 that is taken up by plants along with ordinary carbon during their normal photosynthesis. Nuclear testing in the atmosphere had by 1961 increased the amount of radiocarbon contained in tree rings by 25 percent, far in excess of the normal trace of radiocarbon produced by the influx of cosmic rays. The three scientists noted that if the Tunguska blast were nuclear, they should expect to find excess radiocarbon in the plant material growing at the time.

To test this prediction the scientists examined tree rings from a 300-year-old Douglas fir which fell in 1951 in the Santa Catalina Mountains near Tucson, Arizona, and also from an ancient oak tree which was cut in 1964 near Los Angeles. They found that the level of radiocarbon in the rings of both trees had jumped by 1 percent from 1908 to 1909.

Traces of radiocarbon are produced in the atmosphere by cosmic rays from space that strike air molecules. The Sun's magnetic field helps screen cosmic rays, but when the Sun is at a low level of activity, such as when few sunspots are seen, its magnetic field declines and more cosmic rays can get through to increase radiocarbon levels. The 1909 reading is within the range of radiocarbon produced naturally at times of low solar activity—except that in 1909 the Sun had just passed a peak of activity, and a high radiocarbon level would not be expected.

The picture is confused by erratic fluctuations of up to 2 percent which exist in the levels of radiocarbon measured from 1873 to 1933 by Cowan, Atluri, and Libby. "Such fluctuations tend to obscure the small effect searched for here and make its value the more uncertain," they cautioned. Yet when compared with the average measured over a surrounding forty-year span, the 1909 radiocarbon level still stood out by 1 percent. While such a 1 percent radiocarbon increase could have been due to local effects, and its timing coincidental, Cowan, Atluri, and Libby calculated that sufficient neutrons to account for it would have been liberated by a 5-megaton nuclear explosion.

This might appear to provide at least equivocal support for the nuclear-explosion theory, were it not for an important double-check made far nearer the blast, where the radiocarbon effects would be expected to be more noticeable. Careful measurements from a tree at

Trondheim, Norway, made by J. C. Lerman, W. G. Mook, and J. C. Vogel of the carbon-14 research unit at the University of Groningen in the Netherlands showed no radiocarbon increase in 1909, but rather a steady decrease around that time. Cowan, Atluri, and Libby's results must, after all, have been due to random effects, and not the Tunguska blast.

Actually, there is direct observational evidence against a nuclear explosion at Tunguska, as was pointed out in 1966 by Robert V. Gentry of Columbia Union College, Takoma Park, Maryland. Gentry, a planetary scientist, noted that the fireball from a nuclear explosion would be expected to last thirty seconds or so, whereas according to eyewitnesses the fireball of the Tunguska object's terminal explosion lasted only a few seconds.

It is also worth asking whether a nuclear power plant would be expected to explode even if it were burning up in the atmosphere. According to nuclear engineering experts, and the example of the Soviet nuclear-powered satellite Cosmos 954 which crashed in Canada in January 1978, the answer is no.

Had there been a nuclear blast at Tunguska, its effects would still be detectable at the site in the form of radioactivity. Much of the erroneous popular lore on this point seems to stem from the fictional writings in which Alexander Kazantsev broached the nuclear-explosion theory. For instance, one of Kazantsev's characters speaks of a man who shortly after examining the blast area died in terrible pain as if from an invisible fire. "It could be nothing other than radioactivity," explains the fictional character. But there are no real-life reports that any such thing happened. The Tungus people did refer to "scabs" that appeared on the backs of their remaining reindeer after the fall, and these scabs have in popular writings been interpreted as signs of radiation damage—ignoring the fact that human eyewitnesses who were near enough to be burned by the heat of the falling object showed no signs of radiation sickness. Most likely, the reindeer scabs were simply heat burns caused by the same flash of thermal energy that set fire to the trees; indeed, in view of the tremendous thermal output of the blast, it would be surprising if such heat burns had not been found.

The possible existence of anomalous radiation at the site was carefully checked by geochemist Kirill Florensky of the Soviet Academy of Sciences, who led expeditions to Tunguska in 1958, 1961, and 1962.

Florensky reported in *Sky & Telescope* in 1963 that the only radioactivity in the trees from the Tunguska area was fallout from atomic bombs, which had been absorbed into the wood. "There is nothing specific in the nature of the radioactivity in the area of the Tunguska fall," he wrote.

Other expeditions had noted an acceleration of forest growth in the devastated area, which some put down to genetic damage from radiation. This was another point that Florensky's party looked at in detail. Biologists concluded that only the normal acceleration of second growth after a fire had taken place. The rapid replacement of plants in a devastated area is a well-known phenomenon in ecology.

Kirill Florensky's article in *Sky & Telescope* was titled "Did a Comet Collide with the Earth in 1908?" Among scientists, the comet theory has always been the front runner; in his article, Florensky said that this viewpoint "is now confirmed." His 1962 expedition used a helicopter to chart and collect the distribution of cosmic dust around the explosion area. For 250 kilometers northwest of the site the scientists traced a narrow tongue of tiny particles in the soil, composed of magnetite (magnetic iron oxide) and glassy droplets of fused rock. The expedition found thousands of examples of metal and silicate particles fused together, indicating that the Tunguska body had not been of uniform composition. A low-density stony composition, probably containing flecks of iron, is believed to be typical of interplanetary debris, particularly meteors ("shooting stars"), which are themselves thought to be dust from comets. The particles that spread northwest from the Tunguska blast are apparently the vaporized remains of the head of the comet, dispersed by winds.

A dramatic illustration of what happens to a loosely compacted object entering the atmosphere came on December 4, 1974, when the brightest fireball ever photographed by astronomers descended over the town of Sumava in southwest Czechoslovakia. This object, calculated to weigh 250 tons, completely disintegrated in less than three seconds. Astronomers estimated that its diameter was equivalent to the length of a London bus, and its average density was 0.2 that of water—a loosely compacted object indeed. According to Dr. Keith Hindley, meteor section director of the British Astronomical Association, the Sumava fireball could have been the fragment of a comet's nucleus—in other words, a smaller example of the Tunguska object.

Florensky noted that the approach of the Tunguska comet would have been invisible since it was moving across the daytime sky. "The collision was neither head-on nor a stern chase," he remarked. "The Tungus object struck the Earth almost squarely on the side."

Even if the Tunguska object had caused nuclear effects, this would still not invalidate the comet hypothesis. Astronomer John Brown of Glasgow University and physicist David Hughes of Sheffield University have shown how radiocarbon might be produced by a comet's entry into the atmosphere as the comet burned up and literally turned the air blue with the heat. Although the hot gas produced by the comet's burn-up would have a temperature of no more than a few million degrees, too low for nuclear reactions, Brown and Hughes note that nuclear-like effects are produced in the gases of solar flares at temperatures not much above those calculated for the Tunguska explosion. Therefore they propose that the impact of a comet with the atmosphere might produce X rays, gamma rays, and highly accelerated electrons and nuclei.

Brown and Hughes calculate that sufficient neutrons to explain the radiocarbon data of Cowan, Atluri, and Libby could be produced if the necessary hot plasma conditions were sustained for at least three seconds. This is a plausible figure when compared with, for instance, the duration of long-lived meteor trails and eyewitness reports of the incoming Tunguska body. The lack of observed nuclear effects at the site must mean that the temperature of the plasma was not sufficiently high; perhaps if the comet had met the Earth head-on the energy of the collision would have been sufficient to produce the nuclear effects predicted by John Brown and David Hughes.

A better understanding of the Tunguska comet's motion through space has been provided by astrophysicist Vasilii Fesenkov of the Soviet Academy of Sciences' meteorite committee, who deduced the comet's orbit from the direction and angle at which it struck the Earth. According to Fesenkov, the comet approached from behind the Sun; it had already rounded the Sun and was moving away again into space when it hit Earth. It was never seen because it was always veiled by the Sun's glare. This configuration recalls that of the comet Mrkos in 1957, a bright comet which was not discovered until it was receding from the Sun and had passed the Earth's orbit.

Material from the Tunguska comet's stubby tail, pointing away from the Sun as do all comet tails, would have provided the bright nights seen immediately after the comet's encounter with Earth. This dust from the tail settled rapidly in the atmosphere, most of it vanishing after a day or so. Fine dust from the disintegration of the nucleus, lifted into the upper atmosphere by the updraft of hot air and distributed by stratospheric winds, would have caused the drop in atmospheric transparency around the world as measured by Charles Greeley Abbot in California. The existence of such abundant dust is not consistent with the spaceship hypothesis.

Fesenkov's calculations showed that the Tunguska comet must have had a very elongated orbit, and that it hit Earth at between 30 and 40 km/sec. He agreed that it was loosely compacted, with a diameter of a few hundred meters and a density considerably less than that of water. Its mass he estimated at a million tons.

One of the most significant pointers to the true nature of the Tunguska object came on the night of March 31, 1965, when the Tunguska event was repeated on a smaller scale over North America. An area of nearly a million square kilometers of the United States and Canada was lit up by the descent of a body which detonated over the towns of Golden and Revelstoke, 400 kilometers southwest of Edmonton, Alberta. Residents of those towns spoke of a "thunderous roar" that rattled and broke windows. Sounds were heard 130 kilometers away, and atmospheric shock waves were recorded in Utah, 1,350 kilometers to the south. Seismic waves received at the University of Alberta were equivalent to those from a medium-size earthquake, and the energy released was estimated as equal to several kilotons of TNT.

Scientists predicted the meteorite's point of impact and set out to look for a crater, much as Leonid Kulik had done in Siberia half a century before. Like him, they were unsuccessful. Scanning the snow-covered ground from the air, the scientists were unable to find traces of the meteorite, or of a crater. Only when investigators went into the area on foot did they find that a strange black dust coated the snow for miles around. Samples of this dirt were scraped up, and found to have the composition of a particularly fragile type of stony meteorite known as a carbonaceous chondrite. Once the snow at the site of the Revelstoke fall had melted, no trace of the meteorite dust remained.

As would be expected of such a fragile, low-density object, the Revelstoke meteorite fragmented at a height of about thirty kilometers, raining thousands of tons of crumbly black dust upon the snow below. Significantly, witnesses to the Tunguska blast described just such a "black rain": carbonaceous chondrite dust from the disintegrated head of a comet?

Clinching evidence for the cometary nature of the Tunguska object came from the results of the latest Soviet expeditions to the site, reported in 1977. Microscopic rocky particles found in the 1908 peat layers have the same composition as cosmic particles collected from the upper atmosphere by rockets. Thousands of tons of this material is estimated to be scattered around the fall area. Along with these particles of rock from space were jagged particles of meteoric iron. The Soviet researchers concluded that the Tunguska object was a comet of carbonaceous chondrite composition. This should be no surprise, for astronomers are finding that a carbonaceous chondrite composition is typical of interplanetary debris, including the heads of comets.

Whereas the Revelstoke meteorite was much smaller than the Tunguska object, the heads of known comets are paradoxically too large to explain the Siberian blast; an object such as Halley's Comet with an estimated mass of 50,000 billion tons could produce an explosion equivalent to more than 1 billion megatons of TNT, sufficient to wipe out England. Cometary impacts could have formed some of the giant craters up to several hundred kilometers across on the Moon. The problem is to find an object of intermediate size which could have caused the Tunguska event.

In 1976 astronomers discovered an object of just such a size. A direct repetition of the Tunguska event was avoided only by hours as a previously unknown asteroid swept past at a distance of 1.2 million kilometers, only three times the distance of the Moon—a hair's breadth by astronomical standards. The near-miss of this asteroid, termed 1976 UA by astronomers, was second only to that of the asteroid Hermes in 1937, which missed Earth by only two Moon distances.

Asteroid 1976 UA has an estimated diameter of a few hundred meters, making it the smallest natural object ever observed in space. This is also the size estimated for the Tunguska comet. Although technically described as an asteroid, 1976 UA belongs to a class of objects believed to be the remnant nuclei of "dead" or degassed comets, and may therefore have a carbonaceous chondrite composition. Its generic relationship with the Tunguska comet is quite clear.

Even at its closest, asteroid 1976 UA was far too faint to be seen without a large telescope. Moving at the same closing speed as that calculated for the Tunguska comet, 40 km/sec, the asteroid would take 8.3 hours to cross a 1.2-million-kilometer gap to hit the Earth. Even if it came out of a perfectly dark sky, it would not be visible to the naked eye until twenty-five minutes before impact. Thus a natural object capable of causing another Tunguska blast could easily sneak up on us unnoticed. Fortunately, astronomers estimate that an object the size of the Tunguska comet hits Earth only once in about every 2,000 years, so we can consider ourselves safe—at least for the time being.

REFERENCES

Baxter, J., and T. Atkins, *The Fire Came By* (London: Macdonald and Jane's, 1975).
Ben-Menahem, A., *Physics of the Earth and Planetary Interiors*, Vol. 11 (1975), p. 61.
Brown, J. C., and D. W. Hughes, *Nature*, Vol. 268 (1977), p. 512.
Cowan, C., C. R. Atluri, and W. F. Libby, *Nature*, Vol. 206 (1965), p. 861.
Feenkov, V. G., *Soviet Astronomy*, Vol. 5 (1962), p. 441.
——— *Soviet Astronomy*, Vol. 10 (1966), p. 195.
Florensky, K. P., *Sky & Telescope*, Vol. 26 (1963), p. 268.
Gentry, R. V., *Nature*, Vol. 211 (1966), p. 1071.
Jackson, A. A., and M. P. Ryan, *Nature*, Vol. 245 (1973), p. 88.
Lerman, J. C., W. G. Mook, and J. C. Vogel, *Nature*, Vol. 216 (1967), p. 990.
Oriti, R. *Griffith Observer*, November 1975.
Ridpath, Ian, *New Scientist*, Vol. 75 (1977), p. 346.
——— *Mercury*, September/October 1977, p. 2.
Zolotov, A. V., *Soviet Physics Doklady*, Vol. 12 (1967), p. 108.
Zotkin, I. M., and M. A. Tsikulin, *Soviet Physics Doklady*, Vol. 11 (1966), p. 183.

15

Betty Hill's Star Map

Do we have a star map of extraterrestrial trade routes? This question arises from the most celebrated of all flying-saucer "contact" stories, familiar to millions through John Fuller's book *The Interrupted Journey*, and also known as the Zeta Reticuli Incident because a star map alleged to have been seen aboard an alien craft has been interpreted as identifying the star Zeta Reticuli as the aliens' home. This reputed star map allows us one of the few opportunities to use scientific knowledge to test the facts of a supposed case of extraterrestrial visitation. Such stories have been remarkably barren of verifiable scientific facts since the time of George Adamski, who told of visits in a Venusian spaceship to see cities on the far side of the Moon in 1954, only a few years before space probes showed that the lunar far side was barren and that Venus was hostile to life.

The case discussed here concerns an American couple, Barney and Betty Hill. One evening in 1961 the Hills were returning to their home in Portsmouth, New Hampshire, after a holiday in Canada when Betty became troubled by what she thought was a strange light in the sky following their car as they passed through the White Mountains in the north of the state. Her sister had seen a UFO, but not one which seemed as close as this! Barney stopped the car to examine the UFO through binoculars. He thought he saw faces staring back at him. Alarmed, he ran back to the car and the couple drove home over back roads to escape the UFO, arriving home two hours later than anticipated.

That was all. What happened subsequently is an example of how a UFO story can grow from the simplest of beginnings. It is also a sobering demonstration of how flaw-ridden the most apparently "watertight" of UFO cases are.

What was the light that Betty Hill saw which sparked off the whole case? She described seeing the Moon with a "star" just below it. Later, she said she saw another, brighter "star," which she identified as the UFO. Although the Hills described the weather as perfectly clear, weather bureau reports show that the sky was in fact more than half covered with thin cirrus clouds. Barney did not at first think that the brighter star was anything unusual, until convinced by the prompting of Betty, who had been watching this object for half an hour.

Various aspects of the case have been analyzed by Robert Sheaffer, an astronomer and computer systems analyst. He wrote to Betty Hill to ask her to sketch the relative positions of the Moon, the "star," and the object she called the "craft." Sheaffer knew that the planets Jupiter and Saturn were the only bright objects near the Moon at that time. Betty Hill's sketch showed the "star" close to the actual position of Saturn, and the "craft" near the position of Jupiter. Sheaffer noted that if an unknown craft had been present, Betty Hill would have seen three objects near the Moon. Since she reported seeing no bright objects other than those she sketched, there could not have been any unusual objects present. Jupiter had presumably been hidden by clouds when she first looked. It is tempting to speculate that the entire case would never have developed but for the clouds.*

The object in the sky seemed to move only when the car was moving, and stopped when the Hills stopped. Of course, if an aerial object really had been following the car, it would not have maintained a constant orientation with respect to the Moon. The moral here is clear: accurate estimations of an object's motion can never be made from a

* In the book *The Interrupted Journey*, the Hills described a cigar-shaped object with flashing red, amber, green, and blue lights that crossed in front of the Moon. It is not clear whether this is one of the many details added later (the book was not published until five years after the incident) or whether, as it sounds, it was the observation of an aircraft. Whatever the case, the object does not feature on Betty's sky sketch sent to Sheaffer.

moving vehicle. In the famous "flying cross" incident some years ago, two policemen in Devon, England, reported being "followed" at treetop height by a brilliant object which turned out to be the planet Venus. I cannot mock, because I have been fooled by just such an effect into watching a "moving" UFO that turned out to be the bright star Aldebaran.

Of course, if this were all there was to it, the story would soon have died, like most other UFO cases. But the incident was clearly preying on Betty Hill's mind. She began reading UFO books and was even concerned that she and Barney might have suffered some kind of radiation damage from the supposed craft. *Ten days* after the event, Betty began having a recurrent dream that she and Barney had not escaped from the UFO after all, but that on the back roads they had been abducted by aliens who took them on board the flying saucer for medical examination.

She described this dream to relatives, friends, and others who were interested, including representatives from the UFO group NICAP (National Investigative Committee on Aerial Phenomena). Through the prompting of others, the Hills came to believe that Betty's dreams represented what really happened during the journey home, which explained why they were two hours late. Betty originally gave the time of the abduction as between midnight and 1 A.M., although she has also quoted other times ranging from 11 P.M. to 3 A.M.

Even so, the case would still not have become so celebrated were it not for the next development. Two years after the alleged abduction, Barney Hill was referred to a Boston psychiatrist, Dr. Benjamin Simon, for personal problems. Barney, a black, was of a nervous disposition, with high blood pressure and an ulcer (he died in 1969); Betty, his second wife, is white, with a strong personality. Betty was a supervisor in the New Hampshire department of welfare; since her retirement in 1976 she has become actively involved in UFOlogy. In late 1963 the two of them visited Dr. Simon, who decided that they both needed help. He put them through a course of hypnosis, in which he was skilled. Not everyone is susceptible to hypnosis, but both the Hills were capable of reaching deep states of hypnosis, which is rare; it further underlines the likelihood of a psychological explanation for the case.

Under hypnosis the Hills separately relived the events of that extraordinary night. Though both agreed on the events leading up to the alleged abduction, only Betty offered a detailed account of the abduction itself. In fact, Dr. Simon concluded that Barney's account was based purely on what he had heard of Betty's dreams. Dr. Simon noted a strong sexual symbolism in Betty's description of supposed events aboard the flying saucer: Betty talked of skin tests and of having a needle inserted into her navel, which caused great pain. Dr. Simon was struck by the fantasy-like character of the story, and concluded that it was a retelling of Betty's dream, which she had by then come to believe. The Hills' testimony under hypnosis has been taken as supporting evidence that they really were abducted and taken aboard a flying saucer, but Dr. Simon himself did not believe this. "It was a dream," he has said. "The abduction did not happen." New evidence suggests that testimony of supposed UFO abductees under hypnosis is not reliable (see next chapter).

There is also the matter of the supposed radar confirmation of the UFO. Betty Hill has claimed that seven different radars tracked the UFO, but there are records of only one anomalous radar echo. This was at Pease Air Force Base, near Portsmouth, where runway approach radar showed an unknown target half a mile away at 2:14 A.M. that morning. The wide-angle airport surveillance radar picked up no such target, and it was never seen visually by the control tower, although they were twice told of it. Birds or even insects can register as false targets on radar, but whatever the cause, it was certainly not the flying saucer Betty Hill claimed to have seen a hundred miles away at a different time.

Of all the experiences Betty reported aboard the alien craft, the one that has assumed most significance is the star map she claimed to have seen on one wall. Betty did not consciously remember the map, but mentioned it during a hypnosis session with Dr. Simon in 1964, three years after the alleged encounter, so that it has as much basis in fact as the rest of the abduction dream. After one hypnosis session she drew the map (fig. 7). Her freehand drawing, on which she made two amendments, showed a selection of stars joined by lines; there were, she said, other stars on the original map which she could not remember clearly and did not draw. The strong lines on the map, she said, represented major trade routes, with the dotted lines signifying expeditions.

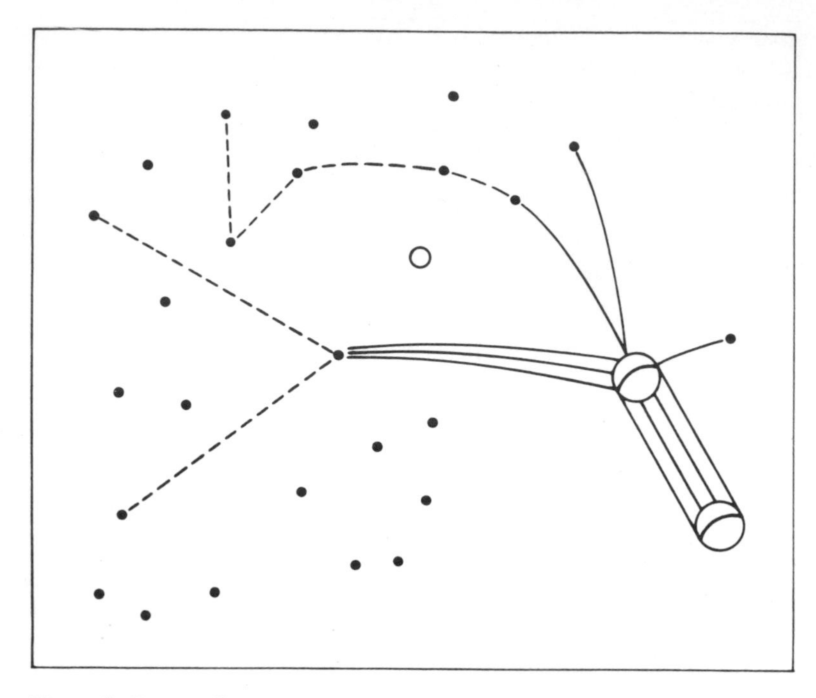

Figure 7. Betty Hill's star map, supposedly seen during her alleged abduction aboard an alien spacecraft.

In 1965 Betty Hill saw in *The New York Times* a star map of the constellation Pegasus including the location of a radio source called CTA-102, which, it was being suggested at that time by Soviet radio astronomers, was a radio beacon for celestial navigation. She immediately interpreted her sketch as a map of the Pegasus region (fig. 8), with CTA-102 acting as a radio beacon to guide the aliens back to their home star, known to earthbound astronomers as Homan, or Zeta Pegasi. This interpretation was reproduced in Fuller's book *The Interrupted Journey*, but its publication proved to be premature; other astronomers soon identified CTA-102 not as an artificial beacon but as a quasar, one of the most distant objects in the Universe, noted for its natural variability at radio and optical wavelengths.

In 1966 Marjorie Fish, a former schoolteacher, now a research as-

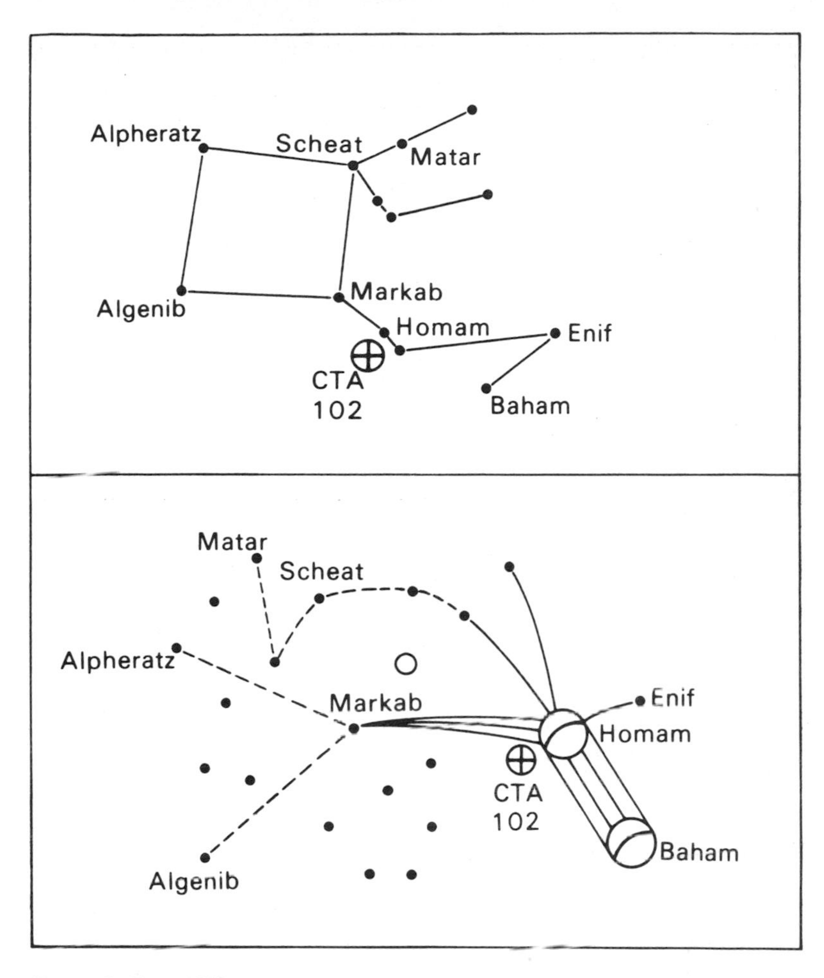

Figure 8. Betty Hill's own interpretation of her star map in terms of the stars of the constellation Pegasus.

sistant with the Oak Ridge National Laboratory in Tennessee, began to investigate stars in the neighborhood of the Sun to see if she could find a fit to Betty Hill's star map. She decided to concentrate on stars like the Sun, because of their presumed suitability for life (see Chapter

2). Not only was the aliens' home star presumably similar to the Sun, but evidently their home planet must have been quite Earth-like, since they reportedly needed no breathing apparatus, and the Hills breathed normally aboard the alien ship.

In 1969 Marjorie Fish visited Betty Hill to learn more about the star map. Betty said that although the map had been flat, with glowing stars, the image on it appeared three-dimensional, like looking through a window. (There had been no mention in *The Interrupted Journey* that the map was supposedly 3-D.) Ms. Fish decided that this newly added detail sounded like a description of a hologram.

By building a model of the solar neighborhood with stars suspended on strings, she was able to select a viewpoint from which a star map like Betty Hill's could be drawn (fig. 9). On this interpretation, the two foreground stars were the two widely separated components of a double star known as Zeta Reticuli. Therefore Zeta Reticuli, 37 light-years away, became the home of the supposed aliens who abducted Betty and Barney Hill. The star itself, barely visible to the naked eye, lies in the southern hemisphere constellation of Reticulum, the Net.

As we saw in Chapter 1, intelligent life arising on a planet of a wide double star would be a prime candidate for developing interstellar travel. Most interestingly of all, astronomical research shows that the Zeta Reticuli pair are each very similar to the Sun in size and brilliance, although they are somewhat older, so that any life would be expected to have reached a more advanced stage. They are separated by about 600 billion kilometers, or 100 times the distance from the Sun to Pluto, so that there would be no theoretical objection to the existence of planets around either star.

Individually, the components are known as Zeta 1 and Zeta 2 Reticuli. The foreground star has been identified as Zeta 2 Reticuli, which is the more Sun-like of the two; whereas Zeta 1 has a luminosity about 70 percent that of the Sun, Zeta 2 has 90 percent the Sun's output. Interestingly, the Hills described conditions as uncomfortably cold aboard the alien craft.

Marjorie Fish's first identification of the star pattern in 1968 was confirmed by the publication in 1969 of the newest and most reliable catalog of nearby stars, compiled by the German astronomer Wilhelm Gliese. However, data from other catalogs, which give different distances for some stars, change the pattern so that it does not fit Betty

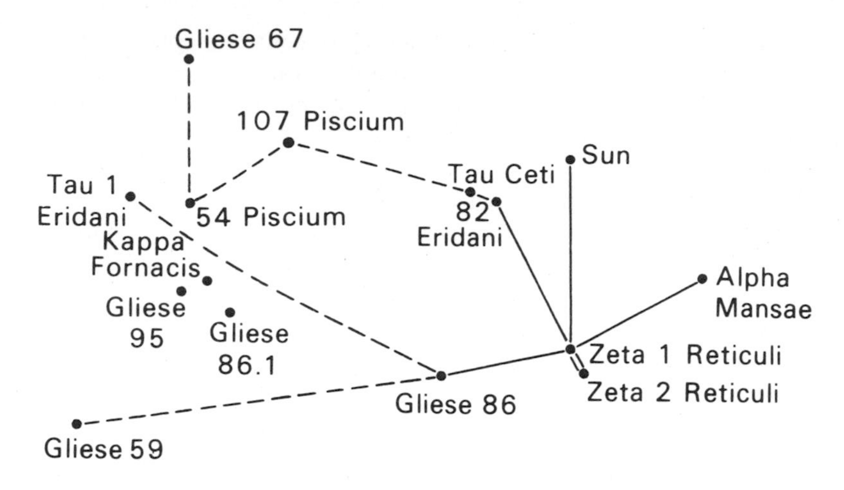

Figure 9. Marjorie Fish's identification of the fifteen main points on Betty Hill's star map.

Hill's sketch. These discrepancies can be clarified only by further research. Neither has it been established how accurately Betty Hill could have reproduced a map from memory after a three-year gap: slight amendments in the "Zeta Reticuli" map could force a major reinterpretation.

The most excitement has been raised by the fact that the stars linked by the "trade route" and "expedition" lines are solar-type stars; but this is, perhaps, less surprising when one remembers that all other stars were eliminated from the search. The Sun itself is at the end of one of the trade-route lines, though what trade is supposed to have been carried on with the solar system by visitors from Zeta Reticuli remains undisclosed.

Although there is no argument with the astronomy behind Marjorie Fish's work, the main points of contention arise when one considers the selection that has been made to find stars that fit the Hill map. For instance, a solar-type star that ought to appear on the map, Zeta Tucanae, is apparently hidden behind Zeta 1 Reticuli. What use is a map that deliberately obscures a major potential target? Equally, other stars in the neighborhood that might be useful for navigation—bright stars such as Sirius and nearby dwarfs—make no appearance on the map. Why not? If they had been included, of course, they would ruin any correlation between the real sky and the Fish-Hill map.

Robert Sheaffer has pointed out that although the Fish map identifies fifteen stars, Betty Hill's original sketch shows twenty-eight points. What about these other stars? Shouldn't they be identified also? There are actually forty-six solar-type stars in the volume of space considered by Marjorie Fish, but only fifteen feature on the map. Why?

Sheaffer notes further that Zeta 1 and Zeta 2 Reticuli are drawn larger than other stars to denote the 3-D effect of the map. Yet, on this assumption, stars such as Tau 1 Eridani and Gliese 95 should also be shown larger, because they are closer than other stars on the map; but they appear only as distant dots. What's more, from the viewing angle chosen, the two components of Zeta Reticuli would actually seem to overlap, although on the map they are drawn widely separated. According to the Gliese data, Zeta 1 and Zeta 2 Reticuli lie at a different angle to each other than shown on the Fish map. Either Zeta Reticuli has been misidentified by Marjorie Fish or the whole map is distorted, in which case all other star identifications are suspect.

Cornell University astronomers Carl Sagan and Steven Soter have pointed out that the only reason Marjorie Fish's star map appears to resemble Betty Hill's original sketch is because of the lines linking the stars. Take away the lines and the similarity disappears (fig. 10). Sagan and Soter argue that if one is allowed to select the stars with which to make the model (as Marjorie Fish did), to select the pattern they are supposed to fit (as happened when choosing fifteen stars out of the twenty-eight on Betty Hill's star-map sketch), and also to choose the viewing point (the star pattern seems to be viewed from an arbitrary point which according to different interpretations lies anywhere between 50 and 200 light-years behind Zeta Reticuli), then it should always be possible to contrive some kind of match. The only problem with this kind of approach is that it is not scientific: the data have been preselected to give a desired result. Using such a pick-and-choose method, one can demonstrate the most impressive correlations between two completely unrelated sets of data. With all the uncertainties associated with the original sketch map, Sagan and Soter concluded that any similarity between Betty Hill's map and the stars is due only to chance.

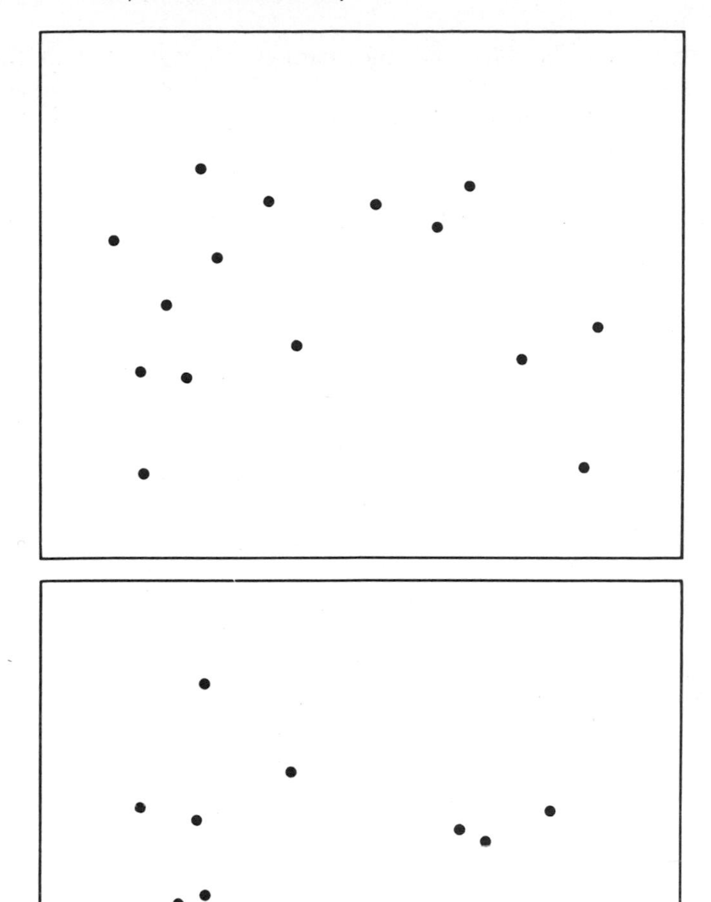

Figure 10. Main points on the Hill map (top) and the Marjorie Fish map (below), when compared without lines linking the stars, show no convincing similarity.

Supporters of the Marjorie Fish map have strongly denied that its claimed resemblance to Betty Hill's sketch could be due to chance. David Saunders, of the University of Chicago's industrial relations center and an active UFOlogist, has calculated that the odds against such a coincidence are more than 1,000 to 1, which in most circles would be regarded as strong confirmation of a proposition.

But such arguments were smashed when it was revealed that Charles Atterberg of Hanover Park, Illinois, a specialist in aeronautical communications, had found an even closer correlation between stars in the solar neighborhood and Betty Hill's map, thus rendering meaningless any statistical claims about the "correctness" of Marjorie Fish's interpretation.

Atterberg's map was completed in 1968, and lay forgotten for many years in the files of UFO organizations. Atterberg, a serious and cautious UFO researcher, decided from the start that all the points drawn by Betty Hill must be regarded as equally reliable, and that it was not valid to ignore some for the sake of convenience. Since there were relatively few stars shown and only a few travel routes, Atterberg concluded that the map must be of a very localized area; this seemed logical, because the UFOnauts would be expected to concentrate their visits on their nearest stellar neighbors. Working on this basis, Atterberg found that from an infinitely distant viewpoint (like that used in flat maps) at coordinates Right Ascension 17h 21.3m, declination −9.1°, a pattern emerged that could be made to fit the Hill drawing. From this location also, Gould's belt, a concentration of distant bright stars in our Galaxy, runs vertically across the map. By contrast, the Fish map has no such independent frame of reference—its orientation is in fact completely arbitrary.

Atterberg identified all but possibly two of the twenty-eight points on Betty Hill's drawing, including such favorites as Barnard's star, Tau Ceti, Epsilon Eridani, and Epsilon Indi, the latter coinciding with the star of origin (fig. 11). Carl Sagan has called these latter stars "the three nearest stars of potential biological interest." Epsilon Indi, 11.2 light-years away, is smaller and cooler than the Sun, so that conditions on an Earth-like planet might well be described as "uncomfortably cold."

Unlike Marjorie Fish, who chose only a selection of solar-type stars to fit on her map, Atterberg chose to work with *all* stars in the solar

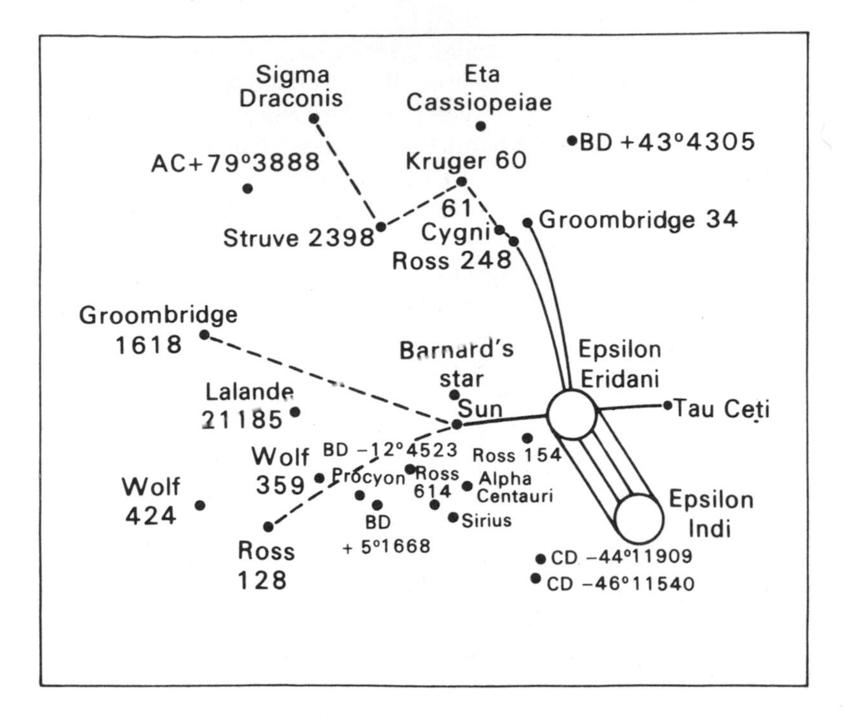

Figure 11. Charles Atterberg's interpretation of Betty Hill's star map, showing tentative identifications of all twenty-eight points.

neighborhood. All the stars marked on the Atterberg map lie within 18.2 light-years of the Sun, which makes the correlation even more remarkable: the Fish map reached out to 53 light-years. The most distant star in the Atterberg map is Sigma Draconis, which appears at the end of the UFOnauts' most extensive expeditionary route.

Although Atterberg's interpretation might at first glance appear to breathe new life into Betty Hill's star map, the opposite is actually true. While the Atterberg map is in many ways superior to the Fish map, it still suffers from some of the same defects. For instance, take away the lines joining the stars and it no more resembles the Hill drawing than does the Fish star pattern. Again, there is no explanation why the two stars of apparent origin should be shown larger than the others; they are not the brightest or the frontmost stars on the map.

Marjorie Fish has criticized the Atterberg map by noting that the UFOnaut trade routes bypass apparently good red-dwarf stars nearby to reach more distant ones; wouldn't the UFOnauts first explore the nearest stars of a given type? Equally, while double stars such as 61 Cygni are joined by lines, other, apparently preferable doubles such as Alpha Centauri are not. Atterberg admits the limitations of any attempted interpretation of Betty Hill's star map when he says: "We are wallowing within a much too wide margin of uncertainty."

The point is that three interpretations of the map have now been offered, starting with the Pegasus version, all of which can claim some compelling individual feature. Given time and ingenuity, doubtless several more "unique" interpretations could be devised, each one exploiting some entirely coincidental factor. The situation is similar to that of the supposed space-probe star map mentioned in Chapter 9.

In short, Betty Hill's star map, although drawn in perfectly good faith, does not turn out to be a reliable guide to the existence of beings from Zeta Reticuli, or anywhere else. As we have seen, there are other disappointing aspects of the "interrupted journey" which indicate that the case is, in fact, less impressive than popularly believed.

Many people will be disturbed by the number of gaping holes and inconsistencies in what is widely regarded as the best-documented of all UFO contact cases. But that is the nature of UFO evidence—a fact which it is as well to understand before embarking on the final chapter of this book.

POSTSCRIPT

Another famous UFO "abduction" case is that of Travis Walton, a woodcutter who claimed to have been taken on board an alien craft in northern Arizona in November 1975, being returned unharmed after five days. According to investigations by Philip Klass, the woodcutting team of which Walton was a member were in danger of incurring a financial penalty for failing to complete a woodcutting contract on schedule, and they staged the UFO "abduction" as an "act of God" excuse for not meeting the contract deadline. A polygraph examiner who tested Walton within a week of the incident concluded that Walton was "attempting to perpetrate a UFO hoax, and that he has not been on any spacecraft." The relevance of the Travis Walton story to this chapter is that only two weeks before the alleged abduction, NBC-TV had shown a program on the Hill case. In August 1977 the Committee for the Scientific Investigation of Claims of the Paranormal urged that the media consult scientists when preparing stories on parascience. The committee singled out the *Reader's Digest* and NBC for special criticism. Should we be surprised if continuing uncritical publicity on the Hill case leads to further fraudulent imitations?

REFERENCES

Atterberg, C. W., *Quest UFO*, 1978 (unpublished).
Dickinson, T., *Astronomy*, December 1974, p. 6.
Fish, M., *Astronomy*, July 1975, p. 41.
Saunders, D. R., *Astronomy*, August 1975, p. 20.
Sheaffer, R., *Astronomy*, July 1975, p. 40.
——— *Official UFO*, August 1976, p. 14, and December 1976, p. 8.
Soter, S., and C. Sagan, *Astronomy*, July 1975, p. 39, and September 1975, p. 16.

16

Visitors from Where?

In recent years the belief that UFOs represent extraterrestrial spacecraft has grown from a minor cult obsession into widespread folk knowledge. Although the term UFO (unidentified flying object) was introduced as a noncommittal alternative to the colloquial term "flying saucer" in a vain attempt at scientific respectability, the two terms have by popular usage become interchangeable and I see no point in trying to distinguish between them. It is a measure of the assumptions inherent in this field that for many people the term "unidentified" is now synonymous with "extraterrestrial spacecraft."

If we were indeed being visited by aliens it would obviously be an important and exciting discovery, particularly in view of the difficulty of finding signs of extraterrestrial life by our own efforts. In principle there is nothing absurd about the idea; but does the suggested evidence stand up to critical analysis? In the latter chapters of this book I have examined several suggested examples of extraterrestrial visitation and found the evidence to be very weak. I have looked in detail at stories of ancient astronauts and examined two celebrated modern UFO cases, finding them to be susceptible to rational explanations. But the extent of the UFO problem is much wider than this. Dedicated UFOlogists will doubtless argue that the problem cannot be dismissed by explaining a few selected examples. So what is the truth about UFOs? Are they indeed the most obvious of all messages from the stars—personal visits by aliens?

That UFOs exist, in the sense that people see things in the sky that they cannot identify, is incontrovertible. Sightings of strange aerial objects go back to the most ancient times, except that then they were described in terms of angels riding celestial chariots. In the 1890s came a rash of sightings of unidentified airships, and nowadays of course they are described as spaceships. The interpretations seem to depend on the state of terrestrial technology. What will they be termed in the next century?

The modern era of UFOlogy began over thirty years ago, on June 24, 1947, when Kenneth Arnold, a thirty-two-year-old businessman piloting a private aircraft saw what he described as a chain of nine "saucer-like" objects that were flying from north to south near Mount Rainier in Washington State. The press eagerly coined the name "flying saucer" for these objects; although no one realized it at the time, this emotive label was to kindle a new mythology. Arnold's report was followed by a flood of other flying-saucer sightings that continues to this day.

Under pressure to explain the apparent invasion of domestic airspace, the perplexed U.S. Air Force set up a series of investigations, the first of which, known as Project Sign, was initiated in 1948. Project Grudge followed in 1949, and in 1952 began the famous Project Blue Book, which was closed in 1969. The Air Force also contracted with the private Rand Corporation for an independent study of the saucer phenomenon.

All the investigations reached essentially the same conclusion: that the so-called saucers presented no threat to national security, that the reports were most probably misinterpretations or fabrications, and that there was no evidence that saucers were extraterrestrial vehicles.

These conclusions did not satisfy committed saucer watchers such as UFO writer Donald Keyhoe, who claimed in the 1950s and 1960s that the Air Force was not telling all it knew. But the official documents, now declassified, reveal that the Air Force knew no more than anyone else. A declassified CIA report of the same period, the Robertson Panel report of 1953, yields no further support for the once popular "cover-up" theory. UFOlogists pricked up their ears again in April 1977 when the *U.S. News & World Report* promised that the saucer-sensitive Carter government would release restricted CIA information by the end of that year. But hopes were soon dashed when it was realized that the report was based on a misunderstanding. President Carter's science adviser Frank Press had asked both the CIA and the Air Force if they still retained UFO information. The answer was no. Carter's next move was to ask NASA if it would be prepared to undertake a new UFO study, a request which the space administration rejected as "wasteful and probably unproductive."

The Air Force explained the Kenneth Arnold sighting as mirages due to a temperature inversion. Arnold had reported that the air at his flying height of 9,500 feet was clear and still—characteristic of such an inversion. The Air Force files also pointed out inconsistencies in Arnold's estimates of the probable sizes and distances of the objects: his attention was drawn to the objects by a series of flashes, but the chances are very slight that a moving object would continually reflect sunlight in a series of flashes to an observer twenty to twenty-five miles away, which is the range Arnold estimated for the objects. Also, at that distance the saucers would have had to be at least 2,000 feet long and moving at supersonic speed to fit his description of their apparent size and motion. The objects, if they were real, were probably smaller, much closer to him, and slower-moving than he proposed. Inconsistencies in a report can indicate that the sighting is a misperception, as was apparently the case here.

An altogether more spectacular case the year following Arnold's sighting sparked off a new sensationalist line: UFOs are hostile. On January 7, 1948, Air Force captain Thomas Mantell was killed while climbing to intercept a high-flying UFO reported by ground control at Godman Field, Kentucky. The position of the reported UFO fits that of the planet Venus, but a Navy Skyhook balloon released that morning was also in the area. These balloons fly at altitudes of 60,000 feet or more, well above the operating height of Mantell's plane, which had no oxygen equipment. Mantell described the object as "metallic, of tremendous size"; his last words, at 15,000 feet, were: "I'm trying to close in for a better look." His crashed plane was subsequently found five miles southwest of Franklin, Kentucky. The Air Force conclusion was that Mantell had blacked out from oxygen starvation during his climb, and the uncontrolled aircraft had spiraled into the ground.

The CIA entered the UFO field at this point, because it was they who were developing the Skyhook program for photoreconnaissance of the Soviet Union. They were both embarrassed that their secret had got out and worried that some UFO cases might actually be sightings of similar Soviet balloons sent to spy on the United States.

In the 1950s came the famous hoax by George Adamski, who claimed he had met extraterrestrial beings from Venus, Mars, and Saturn who had given him rides in their spacecraft to see people living on the far side of the Moon. Adamski's photographs of the reputed flying saucer that picked him up are among the most blatantly fraudulent ever produced, even in a field that has become notorious for trickery: one can even see the multiple reflections from the artificial lights used to illuminate the saucer model. These excesses gave UFOlogists at least one clear lesson: never trust a flying-saucer photograph.

Adamski and his ilk brought the flying-saucer field into ridicule. But it was soon revived by the birth of the space age, first with the launching of Sputnik 1 in 1957 and then Yuri Gagarin's pioneer manned flight in 1961. Here was a practical demonstration that beings can travel through space in rocket ships. Naturally, it also brought reports that astronauts had encountered UFOs, beginning with the "fireflies" seen by John Glenn. These were simple flakes of paint or liquid droplets from the craft itself, but others proved more puzzling.

Astronaut UFOs have recently been investigated by James Oberg of NASA's Johnson Space Center in Houston, Texas. By reference to original NASA photo files, Oberg was able to explain several UFO photographs as long-range views of rocket stages, in one case tracking lights on a rocket stage seen on the night side of Earth, several examples of lens flare, and even a deliberate forgery by a UFO hoaxer who had airbrushed out part of a Gemini spacecraft, leaving two reflections of sunlight from thrusters on the spacecraft's nose appearing like mysterious lights in space.

In 1975 the UFO group NICAP selected a photograph taken by Gemini 4 astronaut James McDivitt as one of the "four best" UFO photographs ever. But they ignored the fact that McDivitt himself explained that the image, a hazy blob with a long tail, was caused by sunlight reflected off a metal bolt onto the dirty window. McDivitt did report seeing in orbit a cylindrical object with an arm, which Air Defense computers could not identify with the known position of any existing satellite. But actually, reports Oberg, the object was the upper stage of the booster which placed Gemini 4 into orbit, with a strap hanging from it. The computer did not take Gemini 4's own booster into account when trying to identify the mystery object.

There are other examples of sightings (and photographs) of satellites and space debris seen either in orbit or, in the case of Apollo 12, on the way to the Moon (these were panels jettisoned from around the lunar module), but there are no signs of any alien spacecraft. The only "evidence" is within the minds of the UFOlogists who, unlike Oberg, have not had the opportunity to check the data for themselves.

Recently there have been suggestions that there are aliens on the Moon, and that NASA has photographic evidence that it is keeping quiet. I was vaguely aware of these rumors but had not paid them much attention until I received some unsolicited information in June 1977 from a religious group in London, quoting a story from a Canadian newspaper. I was surprised to find that their main piece of "evidence" was ten years old. It was a photograph taken by Lunar Orbiter 5 in 1967, which I well remember from my days in lunar research at the University of London Observatory. The photograph shows rocks rolling down the central peak of the lunar crater Vitello, although the UFO group claimed that these were mining machines and that the photograph was still on the secret list. In fact, it was published on page 217 of the October 1967 issue of the world's leading astronomy magazine, *Sky & Telescope*. Rather than being kept secret, the photograph is listed in the current NASA catalog—order photograph number 67-H-1135.

As further "evidence," the UFOlogists claimed that Apollo V had photographed mining machines on the far side of the Moon in January 1969. Actually, Apollo 5 was launched in January 1968, not 1969; it was an unmanned test flight into Earth orbit, and never went anywhere near the Moon. Of course, Apollo 11 landed on the Moon in 1969, but that was in July. In January 1969 there were no Apollo or Lunar Orbiter craft around the Moon.

Now that scientists have become so interested in the possibility of extraterrestrial life, UFO events are being subjected to far more rigorous examination—with the result that many of the most cherished cases are crashing to the ground. Technical journalist Philip Klass of *Aviation Week and Space Technology* magazine has disposed of most of the classic cases of UFOlogy. Among them is the celebrated "landing" at Socorro, New Mexico, in April 1964, which UFOlogists once voted the most impressive case on record. According to the story, patrolman Lonnie Zamora was chasing a speeding motorist toward the edge of town when his attention was drawn by a roaring sound and a bright flame in the desert three-quarters of a mile away. Abandoning his chase, Zamora drove over to what was apparently a landed spacecraft, with two white-clad figures standing nearby. As he approached, the craft took off and disappeared over the nearby mountains. At the site of the purported landing were four pad prints.

Klass found numerous inconsistencies in this story. Despite the loud noise and brilliant flame which Zamora said drew his attention, a couple living a few hundred yards from the site had noticed nothing unusual. Photographs show that despite the reported "intense flame" under the craft, there was no more singeing of a bush and a clump of grass than could have been produced with a cigarette lighter. The reputed pad prints were spaced irregularly; one looked as though it had been formed by moving a rock, while another appeared to have been dug by a small shovel. No one other than Zamora noticed the noisy, flaming craft depart over U.S. Highway 60, although a gas station attendant reported that a customer told him he had seen a police car going out after a flying craft that was coming in to land. But Zamora said he had not driven over the craft until it was already on the ground.

Such blatant inconsistencies point not merely to a misperception, but to a hoax. Klass noted that, by curious coincidence, the local mayor, who is also the town banker, happened to own the site on which the alleged UFO landed. Socorro gained considerable publicity from the case, with an inevitable improvement in the tourist trade.

No more reliable is the case of two shipyard workers at Pascagoula, Mississippi, who claimed in 1973 to have been abducted by UFO occupants with lobster-like claws while fishing from a pier in the Pascagoula River. I have seen that pier, which is only a few hundred yards from the busy U.S. Highway 90 between Mobile and New Orleans, and can confirm that no glowing UFO could have landed without being seen by passing motorists. Yet there were no other reported sightings of this UFO. Klass found that a lie-detector test, the only independent support for the fantastic story, was administered by an unqualified operator under uncontrolled conditions.

Some astounding experiments, reported in 1977, throw a harsh new light on the analysis of such "contactee" cases. The Pascagoula shipyard workers were examined under hypnosis, a technique used increasingly to extract details of UFO encounters, as happened with the Betty Hill and Travis Walton cases reported in Chapter 15. UFO researchers term such abductions "close encounters of the third kind," a classification originated by astronomer J. Allen Hynek and since adopted as the title of a famous film. Some people regard hypnosis as a truth test in itself; alas, experiments by Professor Alvin H. Lawson of California State University, Long Beach, reveal that this is not so.

I received a copy of Professor Lawson's report after an article of mine criticizing UFOlogists had appeared in *New Scientist;* at the time of writing, his report remains unpublished. With the help of Dr. William C. McCall, a highly experienced medical hypnotist who has examined about twenty people supposedly involved in close encounters of the third kind, a group of test subjects were hypnotized and asked to describe imaginary UFO abductions. The subjects had been specially chosen because they had no particular knowledge of the UFO field, nor had they been involved in any UFO cases. Yet the details they related under hypnosis correlated closely with those described by "real" UFO abductees.

In particular, the test subjects recounted how they were taken on board the UFO, examined by strange creatures, and then released unharmed. References to bright lights inside the UFO, strange writing on the wall, uncomfortably cold temperatures, telepathic communication with the humanoid UFOnauts, and subsequent amnesia or even time lapse are all familiar themes in UFO abduction cases. Lawson concluded that the details given by alleged UFO abductees come not from reality but from the witnesses' imaginations. Therefore it seems that we can sweep away all close encounters of the third kind.

Professor Lawson left open the question of what stimulates the abduction fantasy. But by reference to well-known cases we can conclude that, in the language of psychology, UFO fantasies are projections onto normal objects which have been misperceived. As so many cases have shown, once witnesses believe they have spotted a UFO their imagination runs riot. An extreme case is Betty Hill's "interrupted journey," in which a misperception of the planet Jupiter apparently triggered an abduction fantasy. Some cases, by contrast, have no specific stimulus; they are deliberate hoaxes. But even these will seem convincing under hypnosis, as Lawson's experiments show. UCLA psychologist Ronald K. Seigel noted in an article in the October 1977 *Scientific American* that hallucinations are similar from one person to another. The psychologist Carl Gustav Jung also drew attention to the similarity between UFOs and dream images and even UFO images in modern paintings.

Although there is clearly a strong psychological component in many UFO cases, it would be wrong to dismiss all sightings as products of the imagination. The bulk of the UFO problem consists of cases far less spectacular than close encounters of the third kind.

In 1973 J. Allen Hynek, former astronomical consultant to Project Blue Book, set up the Center for UFO Studies in Evanston, Illinois, to coordinate scientific investigation of UFO reports. UFO cases are phoned in to the Center from police departments and other official agencies over a toll-free hot line, which is manned twenty-four hours a day. In November 1976 the Center began publishing a monthly bulletin, the *International UFO Reporter,* which reports on investigation of cases referred to the Center and selects an occasional case of notable merit. The bulletin was perhaps a little unlucky in the choice of its first high-merit case, concerning a silver saucer reportedly seen by fourteen boys during a hike at a summer camp near Winsted, Connecticut. Alas, there were no independent reports of the object, and a letter in the local paper the following week purporting to come from four of the campers confessed it to be a hoax. Nonetheless, the UFO Center rated its chances of being a "genuine UFO" as 50 percent.

In a typical month, March 1977, the Center received 82 reports at an average of 2.6 per day. Of these reports, 95 percent or 78 cases, were identified, the main culprits being bright stars and planets, aircraft, meteors, and satellites. The remaining four sightings were classified as UFOs of limited merit, which means that they remained unidentified after preliminary investigation, but for various reasons were not considered sufficiently reliable to warrant further scrutiny (here the term "unidentified" really does mean what it says!). There was no case of outstanding merit that month.

It hardly needs to be emphasized that throughout the thirty years or more of UFO investigation no one has yet come up with one confirmed case of extraterrestrial visitation. We would certainly have heard all about it if they had! There are many tall tales that would be of great importance if they were true, but investigation has shown that they are not. After twelve years of fairly consistent interest in this field, I do not know of one UFO that has been identified as an extraterrestrial spacecraft—although I know of many sightings that have been identified as natural or man-made objects.

When something unexpected but undeniably real does occur, there is usually no shortage of witnesses or confirming evidence. For instance, on the afternoon of August 10, 1972, a gigantic fireball streaked through the atmosphere, missing the Earth's surface by forty miles before speeding out into space again. Such an event is exceptionally unlikely, but nevertheless it was seen by thousands of eyewitnesses, including a meteor expert, and was extensively photographed, even though it passed over one of the most sparsely inhabited regions of North America, from Utah to Alberta. It was also detected by a U.S. Air Force surveillance satellite.*

As another example, a bright meteor that split into two as it entered the atmosphere over California on March 22, 1977, produced reports to the Center for UFO Studies from six sheriffs departments, eight police officers, and nine pilots at three separate airports, even though the event occurred at 3:30 A.M. On December 21, 1968, I well remember seeing a peculiar, starlike object in the western evening sky surrounded by a fuzzy glow—a "classic" UFO. Telephone calls quickly confirmed that many fellow amateur astronomers in Britain had spotted it too; several of them photographed it. The effect we were watching was caused by a fuel dump from the third stage of the rocket that had boosted Apollo 8 toward the Moon earlier that day.

Most UFO reports (though not all) are made by people who are unfamiliar with the sky, and are genuinely baffled by something they are unable to identify. Even Jimmy Carter's famous UFO sighting, made in 1969 but not revealed until he was running for President, turned out to have been a misidentification of the planet Venus. Unlike the three cases of unexpected but actual effects reported above which were widely seen and photographed, a disproportionate number of UFO reports seem to have limited witnesses (often only one) and no reliable photographic evidence. Such UFOs seem to be ephemeral: they can apparently materialize and dematerialize at will and have a wide range

* For photographs and further details of this remarkable event, see *Sky and Telescope*, July 1974.

of abilities, such as sudden turns and acceleration and supersonic travel without a sonic boom. It seems, in fact, that UFOs have whatever attributes we wish to assign them, including the ability to avoid conclusive detection. UFO skeptic Robert Sheaffer has noted that theories about such dubious phenomena are set up so that they can never be refuted, no matter how often they fail. Since these theories forbid nothing, they tell us nothing, and hence have no scientific value.

From this I deduce what I term the UFO Uncertainty Principle: one cannot have a UFO sighting which is both highly reliable and highly specific. By a reliable sighting I mean one with many independent witnesses who all agree on what they have seen; when these occur, the object is either something like an unspecific light in the sky, or it is rapidly identifiable, as in the cases mentioned above. By a highly specific sighting I mean one which speaks of such things as silvery craft and alien occupants; but these seem to occur only to isolated individuals or small groups, and are highly unreliable, as in the Betty Hill, Socorro, and Pascagoula cases.

Mathematician Paul Davies of King's College, London, notes that over the past fifty years a new category of event called ball lightning has emerged from reports of strange aerial phenomena. Reports by the public and scientists of luminous objects that move erratically and disappear by exploding are now categorized by this name, even though its true nature is not understood. Davies argues, reasonably enough, that among UFO reports may be evidence of other, still unknown atmospheric phenomena. But if a previously unexpected phenomenon such as ball lightning can be identified and become accepted, why cannot an expected phenomenon like alien spacecraft? One compelling answer is that such craft do not exist. Indeed, there are so many stimuli, natural and man-made, in the sky that we do not even need the extra hypothesis of alien visitation to explain UFO reports.

This should not be surprising, since the scientific assessment of extraterrestrial life discussed in this book implies that not one of the hundreds of thousands of UFO reports on file throughout the world represents an extraterrestrial visitation. Whereas UFOlogists have always argued that once the obviously erroneous sightings have been weeded out there remains a so-called residue of apparently baffling cases, the dedicated researches by modern UFO investigators have shown that even the supposedly "best" of these cases is soluble.

I believe that investigation of UFO reports should continue, at least for a while, and their explanations be widely publicized, because to the general public and the credulous UFOlogists every sighting that remains unexplained becomes one more example of extraterrestrial visitation. I am concerned that the wide sensationalism of UFO stories by the media is leading to a public hysteria in which fantasized details are projected onto ordinary objects such as planets that have been misperceived, and that even entire encounters are being fantasized. With ever more alleged cases, and ever more amateur UFO research groups being formed to look into them, UFOlogy is facing a crunch of having to deliver the goods.

CNES, the French space agency, began an official UFO study in 1977, the same year that NASA rejected President Carter's request to begin a study of its own to supersede the controversial Condon Report of 1969. Only a study which solicits the views of believers and skeptics alike will command the respect of all. Once it has swept aside the smokescreen of misinformation that cloaks this whole subject, I doubt that such a study will reach a different conclusion from those before it. But this time everyone will have to accept it.

I have already argued, in Chapter 10, that the best way of making first contact is by radio, and so far there has been no sign of any such radio signals. The detection of real starships was discussed at the end of Chapter 6; so far, no starships answering to those descriptions have been observed. Yet, according to what I read from UFOlogists, we are under continuous, exhaustive survey by many types of craft and, seemingly, many different beings, from some unspecified base or bases. The purpose of this scrutiny is not clear, nor is the need for such frequent visits. Such a belief builds on the basic fallacy that we are important enough for other people to be deeply interested in us. If there are a lot of other civilizations out there, then we certainly are *not* important. And if there are not a lot of civilizations out there, where are all these visitors coming from?

The UFOnauts described in so many flying-saucer stories seem to be incredibly dumb for creatures apparently capable of interstellar travel—witness their numerous reported landing goofs, their clumsy medical examinations that seem to be *de rigueur* for all contactees, and their general inability to communicate with people. No Earth society would fund such an inefficient bunch of space explorers.

Imagine that we discovered life on another planet. We would either land and make our presence known, or we would observe circumspectly, from a distance, so that no one would know we were there. Unless interstellar travel is *much* easier than we have imagined, we would not keep making frequent visits. We might resurvey the planet every few centuries, but in between we would leave a machine to keep watch for us, like the Bracewell probes described in Chapter 9.

An increasingly popular theory is that UFOs are not spacecraft from other worlds, but visitors from another dimension—perhaps time travelers. But changing the theory does not change the lack of factual evidence. Despite more than thirty years of study, the field of UFOlogy has failed to produce *one* concrete example of visitation, from any dimension. Most scientists would draw their own conclusions from such an abject lack of results, but they do not have the indefatigable optimism of committed UFOlogists for whom the Perfect Case, like the Second Coming, is an article of faith.

The basic appeal of UFOlogy for the masses is that it is a belief system, rather than a field of scientific investigation. This explains why scientists engaged in the search for extraterrestrial life are not besieged by the millions of readers of UFO books. People prefer their dreams. UFOlogy loses its value as an escape mechanism when examined too closely.

In August 1977 the Committee for the Scientific Investigation of Claims of the Paranormal attacked the media for its growing uncritical coverage of parascience. If similar distorted treatment were given to politics or history, there would be a major outcry. While some newspaper and TV companies lay out considerable sums on credulous items about UFOlogy and related topics of parascience, similar interest is not shown in investigative journalism reporting rational solutions to mysteries and exposing the hoaxes and frauds.

The power of the media in promoting misinformation in this field is not to be underestimated. As an April Fool's joke in 1977, Anglia TV in Britain broadcast a widely publicized spoof documentary which invented a fantastic story about a space "cover-up" and a secret U.S.-Soviet manned landing on Mars. In this case, no blame attaches to the TV company, which was quite open about the leg-pull and even dated the program credits April 1.* Yet I still found myself arguing months later with members of a UFO group who told me that the TV company wouldn't have gone to those lengths unless there were "something in it." The "documentary" was a telling example that no matter how absurd the tale or however unconvincingly it is presented, there will always be a small percentage of people who believe it.

By the same token, many people believe that there would not be such a fuss about UFOs unless there were "something in it." For those who wish to cash in on their belief, I mentioned Ladbroke's UFO bet in Chapter 1. The sensationalist *National Enquirer* has upped its reward for proof that UFOs come from outer space and are not natural phenomena from $100,000 to $1 million, with a $10,000 prize for each year's best UFO story (perhaps this is one reason why the number of elaborate hoaxes is on the increase). And in 1978 the Cutty Sark scotch whisky company in the UK offered 1 million pounds to anyone who finds an alien spaceship or part of one.

UFO skeptic Philip Klass has demonstrated his confidence that there are no spaceships from other worlds in our skies with a $10,000 wager. According to the terms of the wager, which has been taken up by three UFOlogists, Klass agrees to pay $10,000 to each of the other parties should one of three events occur: when a spacecraft or fragment is found that is clearly of extraterrestrial origin, in the opinion of the U.S. National Academy of Sciences; when the National Academy of Sciences announces other information which it believes constitutes proof of extraterrestrial visitation in the twentieth century; or when the first alien visitor appears before the United Nations or on a national TV program. The other signatories to this wager each undertake to pay Klass $100 each year that none of these events occurs, for up to ten years. Once the ten years are up, their payments cease, although Klass's obligation remains in force until his death or that of the signatories. Also, should any of the specified events come to pass, Klass has offered to refund the purchase price of his book *UFOs Explained* to all those disgruntled readers who return their copies.

The day that Klass, the *National Enquirer*, Cutty Sark, or Lad-

* Erich von Daniken, incidentally, claims that *New Scientist* magazine supports his ancient-astronauts viewpoint, citing an article in that magazine which was deliberately published on April 1.

broke's pays out, I will be happy to admit that the first extrate
visit has finally occurred.

REFERENCES

Davies, P. C. W., *Theoria to Theory*, Vol. 9 (1975), p. 107.
Klass, P. J., *UFOs Explained* (New York: Random House, 1974; Vintage Books, 1976).
Oberg, J. E., *Official UFO*, October 1976, p. 12.
Ridpath, Ian, *New Scientist*, Vol. 75 (1977), p. 77.
——— *Official UFO*, September 1977, p. 20.
Steiger, B. (ed.), *Project Blue Book* (New York: Ballantine Books, 1976).

Index

Page numbers in italic refer to illustrations.